TEACHER GUIDE
4th–6th Grade

Includes: ...edule | ...W... Activities | S... Tests

Elementary Anatomy: Nervous, Respiratory, & Circulatory Systems

Author: Lainna Callentine

Master Books Creative Team:

Editor: Laura Welch

Design: Terry White

Cover Design: Diana Bogardus

Copy Editors:
Judy Lewis
Willow Meek

Curriculum Review:
Kristen Pratt
Laura Welch
Diana Bogardus

First printing: April 2016
Eighth printing: October 2022

Copyright © 2016 by Lainna Callentine and Master Books®. All rights reserved. No part of this book may be reproduced, copied, broadcast, stored, or shared in any form whatsoever without written permission from the publisher, except in the case of brief quotations in articles and reviews. For information write:

Master Books®, P.O. Box 726, Green Forest, AR 72638
Master Books® is a division of the New Leaf Publishing Group, Inc.

ISBN: 978-1-68344-319-3
ISBN: 978-1-61458-824-5 (digital)

Unless otherwise indicated, Scriptures taken from the Holy Bible, New International Version®, NIV®. Copyright © 1973, 1978, 1984, 2011 by Biblica, Inc.™ Used by permission of Zondervan. All rights reserved worldwide.

Scripture quotations marked (NLT) are taken from the Holy Bible, New Living Translation, copyright ©1996, 2004, 2007, 2013, 2015 by Tyndale House Foundation. Used by permission of Tyndale House Publishers, Inc., Carol Stream, Illinois 60188. All rights reserved.

Printed in the United States of America

Please visit our website for other great titles:
www.masterbooks.com

Permission is granted for copies of reproducible pages from this text to be made for use with immediate family members living in the same household. However, no part of this book may be reproduced, copied, broadcast, stored, or shared in any form beyond this use. Permission for any other use of the material must be requested by email from the publisher at info@nlpg.com.

About the Author: Dr. Lainna Callentine, MEd, MD, is a physician, instructor, writer, speaker, and creator at Sciexperience and volunteers her services at a clinic that serves the uninsured in the Chicago suburbs. She is a coach, teacher, pediatrician, and homeschool mother. She obtained her bachelor's degree in human development and social policy from Northwestern University. She obtained her master's degree in education from Widener University. Then she went on to pursue her lifelong dream of becoming a doctor. Dr. Callentine obtained her medical degree from University of Illinois College of Medicine.

> Your reputation as a publisher is stellar. It is a blessing knowing anything I purchase from you is going to be worth every penny!
> —Cheri ★★★★★

> Last year we found Master Books and it has made a HUGE difference.
> —Melanie ★★★★★

> We love Master Books and the way it's set up for easy planning!
> —Melissa ★★★★★

> You have done a great job. MASTER BOOKS ROCKS!
> —Stephanie ★★★★★

> Physically high-quality, Biblically faithful, and well-written.
> —Danika ★★★★★

> Best books ever. Their illustrations are captivating and content amazing!
> —Kathy ★★★★★

Affordable
Flexible
Faith Building

Table of Contents

Using This Teacher Guide 4
Course Objectives 4
Course Description 5
Eight Areas of Intelligence 6
Suggested Daily Schedule 9

The Electrifying Nervous System
    Nervous System Objectives 16
    Supply List for Activities 17
    Activities and Worksheets 19

The Breathtaking Respiratory System
    Respiratory System Objectives 129
    Supply List for Activities 131
    Activities and Worksheets 133

The Complex Circulatory System
    Circulatory System Objectives 231
    Supply List for Activities 233
    Activities and Worksheets 235

Exams 359
Portfolio/Rubrics/Reports 377
    A. Biography Rubric
    B. Oral Report Rubric
    C. Science Experiment Rubric
    D. Objective Concept for Nervous System
    E. Objective Concept for Respiratory System
    F. Objective Concept for Circulatory System
    G. Scientific Report Form
    H. Learning Log

Answer Keys 399
Bibliography 418

Using This Teacher Guide

Features: The suggested weekly schedule enclosed has easy-to-manage lessons that guide the reading, worksheets, and all assessments. The pages of this guide are perforated and three-hole punched so materials are easy to tear out, hand out, grade, and store. Teachers are encouraged to adjust the schedule and materials needed in order to best work within their unique educational program. **There is a supply list for the activities in this teacher guide at the front of each unit study.**

God's Wondrous Machine! Go on an amazing journey through your body's nervous, respiratory, and circulatory systems! Students will learn how their brains control the different parts of their body, how the cycle of a breath works, what route their blood takes through their body, and much more. Providing a variety of worksheets that appeal to different learning styles and skill levels, this course is instructional and exciting for any student.

🕐	Approximately 30 to 45 minutes per lesson, five days a week
🔑	Includes answer keys for worksheets and tests
✏️	Worksheets for each section
✓✏️	Tests are included to help reinforce learning and provide assessment opportunities
🔄	Designed for grades 4 to 6 in a one-year science course

Course Objectives: Students completing this course will

- Investigate the main areas and structures of the brain and what important role each plays in making the body function
- Evaluate awesome examples of God's creativity in both the design and precision of human anatomy
- Review a timeline of important discoveries and innovators, as well as key anatomical terms and concepts
- Explore the human body's respiratory system, focused on structures, function, diseases, and God's efficient and effective designs
- Learn about the mechanics of the circulatory system, how it transports nutrients, blood, chemicals, and more to cells within the body
- Identify important innovations that help professionals understand the mechanisms of our lungs, sinus cavities, and diaphragm
- Demonstrate vital facts about why you sleep, what foods can superpower your brain's functions, and how it controls the wondrous machine known as your body!

Course Description

This series delights in sharing the truth to children of how they are wonderfully made! Beyond the basics of how and why the body works as it does, it is important to share how the amazing and deliberate design of their bodies enables it to function as it should, just as God meant for it to. Utilizing *God's Wondrous Machine* by pediatrician and instructor Dr. Lainna Callentine, students will learn about the complex circulatory system, the electrifying nervous system, and the breathtaking respiratory system, with features that include instructional guidance on the eight areas of intelligence to help students of all learning styles. This includes designated levels and pacing suggestions, and it should be noted that all activities can be used at any level.

Each of the activities and worksheets in this guide have been identified by the various learning styles. Many of these activities can be designated in multiple categories. Remember this is just a guide. The activities can be designated in other ways. If you would like to know which multiple intelligence type a particular activity sheet or worksheet was designed for, you can check in the appendices for the Activity/Worksheet Overview charts.

Note: Keep your worksheets in a folder to have them ready for your review.

Eight Areas of Intelligence

Let's face it. We all learn in different ways. I may be naturally talented in playing basketball. Any sport that I pick up I achieve good success . . . however, I can't carry a musical tune. In fact, I believe people would pay me *not* to sing. We all have different talents with which God has blessed us. Some things come easier than other things. As a former classroom teacher, coach, pediatrician, and homeschool mother, I have witnessed the many talents and ways that my students, players, patients, and children are gifted.

We all are gifted. God places those gifts in each of us. Although I was able to meet with a moderate amount of educational success in my formative years, it has been thwarted by many challenges. My teachers did not appreciate my particular learning style. I was not a traditional learner. Just reading a book and doing worksheets never seemed to help me gain a firm grasp on my studies. I learned best by movement, experiencing, and visualizing my lessons. I see the world in pictures. My constant doodling in class was at times not embraced by my instructors. In fact, it was viewed as a distraction and inattentiveness. This is how I learn. All through medical school, I had the "best" illustrated notes. Even to this day, during Sunday morning sermons I take artistic renditions of the pastor's message. It is through my illustrations that I understand and process what is being said to me.

How effectively we process new information determines how successfully we are able to recall that same knowledge later. The layout of this series capitalizes on hands-on activities, experiments, worksheets, and fascinating stories connecting the student to information engaging the many learning styles of children. Educational trends today focus on linguistic and mathematical abilities almost exclusively. The theory of multiple intelligences was constructed by a developmental psychologist named Dr. Howard Gardner. He is a prolific author in educational theory. His most noted work, *Frames of Mind: The Theory of Multiple Intelligences*, suggested that there are at least eight different types of human intelligence or ways of understanding the world around us. In his book, he discusses how most individuals rely on one or two dominant intelligences. In our quest to acquire knowledge to understand our Heavenly Father and the world that lies around us, it is important to strengthen all of our levels of intelligence.

The eight areas of intelligence are the following:

 INTRAPERSONAL

 VERBAL-LINGUISTIC

 VISUAL-SPATIAL

 MUSICAL

 BODY-KINESTHETIC

 INTERPERSONAL

 LOGICAL-MATH

 NATURALIST

It can be very rewarding to capture your student's interest based on his or her particular learning style and then stretch him or her to develop skills in the other intelligences. God calls us at times to step out of our comfort zone. The more we follow Him and allow that discomfort to occur . . . the more He can use us.

INTRAPERSONAL	VERBAL-LINGUISTIC	VISUAL-SPATIAL
These are the people who are introspective. They tend to understand themselves well. They analyze their thoughts and feelings. They enjoy individual activities. They are "self wise."	These are the people who love to color the world through their words. They think in words. They learn best by writing, reading, and speaking. They are "word wise."	These are the people who think in shapes, colors, and images. They can see the spatial relations in things and know that things will fit just by playing with them in their minds. They are "picture wise."
MUSICAL	LEVEL	BODY-KINESTHETIC
These are the people who can pick up a tune naturally. They hear it once and instantly "get it." They are aware of rhythms and learn best with activities that involve music. They are "music wise."		These people have good physical awareness. They can bound on the playground from apparatus to apparatus like a billy goat scaling the side of a mountain. They are the ones who need to move, and they benefit best through hands-on discovery. They are "body wise."
INTERPERSONAL	LOGICAL-MATH	NATURALIST
These people enjoy working in groups and playing on teams. They enjoy their experiences best with others. They are "people wise."	These people are rational intellectuals. They can see the abstract. They work best with numbers of patterns. They are "logic wise."	These people are acutely aware of the many patterns in nature. They learn best when activities involve animals, plants, and the outdoors. They are "nature wise."

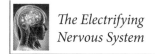

The Electrifying Nervous System | After Pages 14–21 | Day 5 | Worksheet 1 | Name

Just the Facts

Match the word with its related meaning:

A. Anatomy   _____ Abnormal health consequences of disease

B. Physiology   _____ Microscopic cell structure

C. Histology   _____ Name and location of parts of the body

D. Pathology   _____ How the body functions

Quick questions:

1. How much does your brain weigh?

2. At what rate can your brain and nervous system send out signals to the body?

3. What basic function does your brain serve to do?

4. The Edwin Smith Surgical Papyrus was written by what ancient culture?

5. Who is also known as "The Father of Medicine"?

6. Who believed the brain was just a place to cool blood from the heart?

7. What does the Latin word *plumbum* mean?

8. Who is the "Father of Anatomy"?

9. What was the study known as phrenology?

10. When was the first documented and successful removal of a brain tumor done?

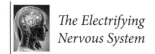

The Electrifying Nervous System | After Pages 14–21 | Day 6 | Activity 2 | Name

Back in Time

Choose one of the people from the historical timeline. Write a short story of how this discovery may have been made — and you can be as creative as you like. For example, imagine a situation that Hippocrates would have felt the need to develop the Hippocratic Oath. Or why the Edwin Smith Surgical Papyrus was written. Or what a day in an early apothecary may have been like.

_____
_____
_____
_____
_____
_____
_____
_____
_____
_____
_____
_____
_____
_____
_____
_____
_____
_____
_____
_____
_____

How Did It Happen?
Short Story Challenge

Imagine you are the assistant of one of the people listed on the timeline of brain-related discoveries or innovations. In 750 words or less, create a possible scenario that might have led to the discovery.

For example, you are Dr. Alice Hamilton's nurse and she is looking over a stack of patient records. When she realizes that the patients all have the same symptoms, she then tries to discover other things they have in common. (Hint: What kind of jobs do they have?)

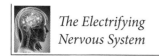

The Electrifying Nervous System | After Pages 14–21 | Day 8 | Worksheet 2 | Name

Biblical References #1

Read the following verses: Psalm 26:2; Matthew 22:37; Colossians 3:2; Psalm 48:9; and Psalm 119:27. Each passage uses the words "mind" and "meditate." Write a short summary of the importance of "mind" and "meditate" as described in these passages.

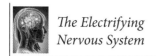

The Electrifying Nervous System | After Pages 14–21 | Day 8 | Worksheet 3 | Name

Biblical References #2

Copy the following verse:

> Test me, LORD, and try me, examine my heart and my mind. (Psalm 26:2)

Write in cursive the following verse: Test me, Lord, and try me, examine my heart and my mind. (Psalm 26:2)

| The Electrifying Nervous System | After Pages 14–21 | Day 9 | Worksheet 4 | Name |

The Word of God

Look up and write Colossians 3:2.

What does this passage mean to you?

The Electrifying Nervous System | After Pages 14–21 | Day 10 | Activity 4 | Name

Supercilious

Here is an intellectual play on words. There is a ridge above the eye sockets in the skull called the *superciliary ridge*. What does it mean when someone describes a person as acting in a *supercilious* way? How are these two terms — superciliary ridge and supercilious — related? Hint: You will need a dictionary!

(CC BY-SA 1.0)

Looking Inside the Brain

Modern technology has afforded us the ability to look into a person's brain. It provides useful diagnostic information to treat disease.

Fill in the blanks below on the following modalities.

1. CT scans or _____ _____ are used for

   diagnosing _____ _____, _____ _____, and

   _____.

2. EEG or _____ is a way of recording _____

   _____ of the _____.

3. MRI, or _____ _____ _____, are _____

   that use powerful magnetic _____ and radio _____ to form images of the

   body.

4. PET scan, or _____ _____ _____, uses

   _____ _____. It reveals which areas of the brain are

   _____.

Techy

MRI (Magnetic Resonance Imaging), CAT (Computerized Tomography), and PET (Positron Emission Tomography) scans have become valuable tools for peering into the body. Physicians use these tools to diagnose problems inside the brain without performing surgery. Write a report describing the difference between these diagnostic tools and describe how they relate to the brain and the nervous system.

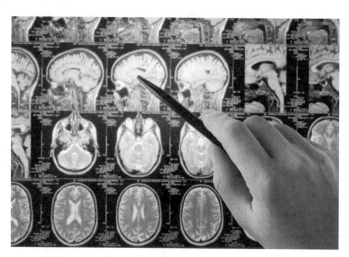

Page left intentionally blank.

The Electrifying Nervous System | After Pages 24–25 | Day 13 | Worksheet 6 | Name

Back to the Basics

Fill in the blanks with the following words:

neuron   dendrites   axon   myelin sheath   neuroglia

The _____ are tentacle-like structures that extend from the cell body and reach out to the other cells.

A long tail-like extension of the cell body is called a(n) _____ and it is surrounded by a white fatty segmented covering called a(n) _____.

Electrical impulses are transmitted through the _____.

_____ literally means "neuron glue."

Match the neuroglia with its function.

_____ Microglia

1. "The grocer" — supplies nutrients to the neuron

_____ Astroglia

2. "The lining" — cells that line the small cavities of the brain and produce cerebral spinal fluid (CSF)

_____ Oligodendroglia

3. "The garbage collector" — these are the phagocytic cells that digest microorganism invaders and waste products from the neurons

_____ Ependymal cells

4. "The protector" — cells that support and insulate the axons by helping to form the myelin sheaths that protect the neuron

Name the parts of a neuron.

1. _____

2. _____

3. _____

Draw and label a picture of a neuron.

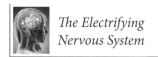

The Electrifying Nervous System | After Pages 24–25 | Day 14 | Worksheet 7 | Name

The Basics of the Nervous System

Fill in the associated boxes.

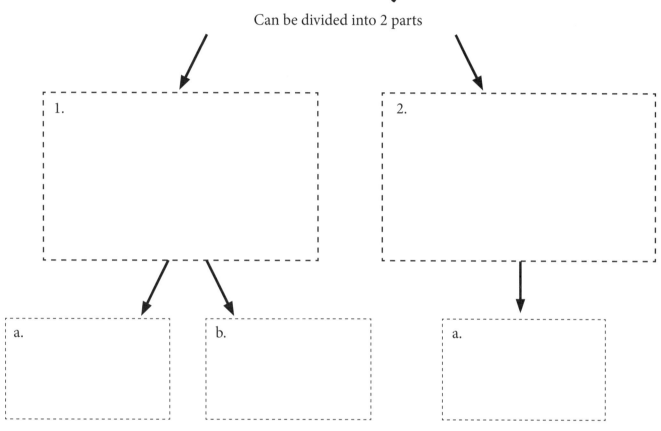

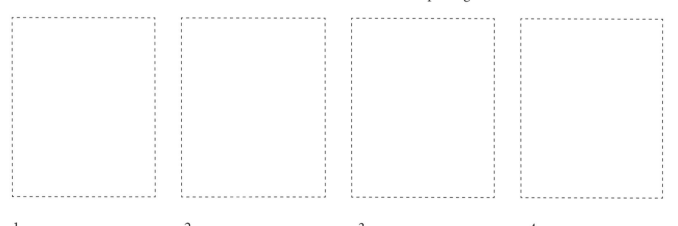

1. _____ 2. _____ 3. _____ 4. _____

Page left intentionally blank.

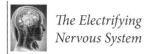

The Electrifying Nervous System | After Pages 24–25 | Day 15 | Activity 6 | Name

Timeline Shuffle

Cut out the following images and paste them in the appropriate spot on the timeline (pages 55–57).

Aristotle

Christ

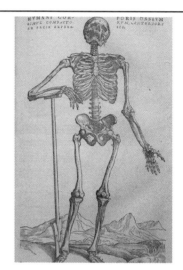

Andreas Vesalius's
De humani corporis fabrica

Dr. Alice Hamiliton

Edwin Smith Papyrus

Dr. Paul Broca

Activity 6, Day 15 // 47

Page left intentionally blank.

Hippocrates

Dr. Raymond Damadian

Dr. Wilder Penfield

Rene Descartes

Galen

Anthony Van Leeuwenhoek

Page left intentionally blank.

Tape the timeline pages together

- 1700 BC
- 460–379 BC
- 335 BC
- 170 BC
- 4 or 5 BC
- AD 1543

Page left intentionally blank.

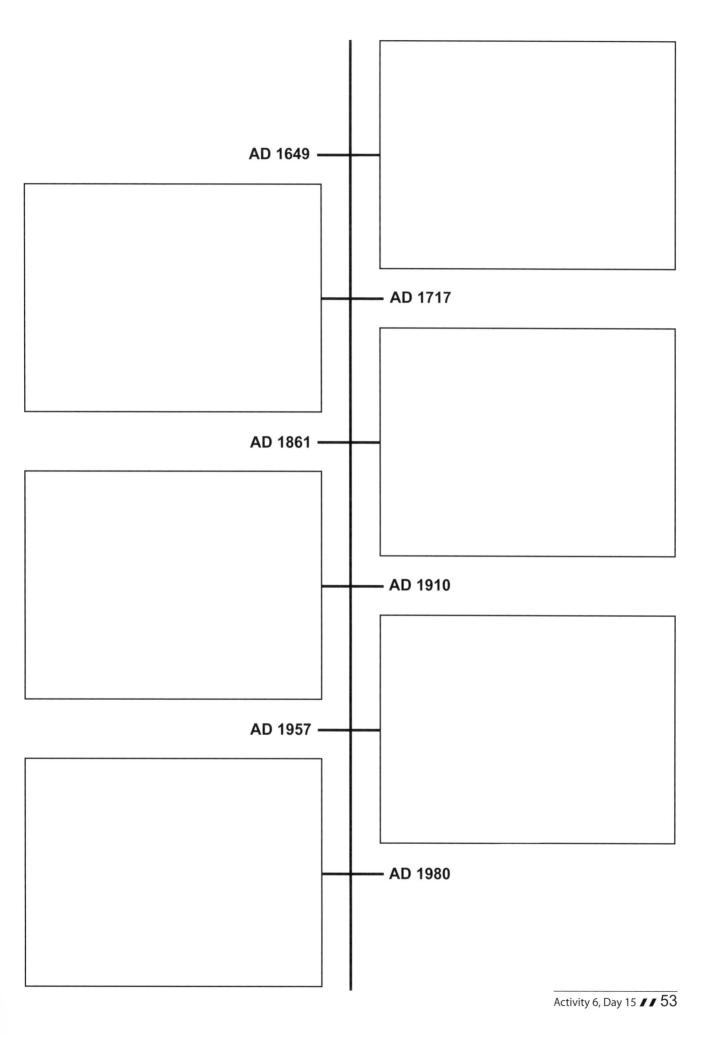

Page left intentionally blank.

| The Electrifying Nervous System | After Pages 24–25 | Day 16 | Activity 7 | Name |

Neuron Connections
Constructing the Super Highway

The brain has more than 15 million nerve cells. The connections that run from every one of these cells relay messages throughout the brain. It is a neurological switchboard.

Choose between the following two projects to illustrate this point:

1. **Materials:** Paper and 10 colored pencils, crayons, or markers

   Instructions: Make 10 dots of different colors anywhere on the paper. Pretend that each of these dots represents nerve cells. Using a red pencil, locate one red dot and draw a line from that dot to each of the other dots. These lines represent the dendrites. Then systematically draw nine lines from the remaining dots to each of the other dots.

OR

2. **Materials:** Scrap piece of wood (measuring at least 12 inches by 12 inches), 10 nails, red yarn, hammer, and scissors

   Instructions: Hammer the 10 nails randomly into the board. Take a piece of the yarn and wrap it around one nail to fix the end. Then take nine separate pieces of yarn and connect one end around the initial nail to each of the remaining nine nails. Do the same process for the other nine nails. The project is complete when you have nine pieces of yarn starting from each nail.

You've Gotta Nerve

The objective is to design a model neuron. Choose from the following ideas.

Possible List of Materials for Each Activity:

1. Modeling clay, cardboard
2. 1 cup of flour, ½ cup of salt, 2 tsp. cream of tartar, salad oil, food coloring
3. Pipe cleaners (5 colors)

Activities:

1. Utilize modeling clay to make a neuron and mount on a stiff piece of cardboard.
2. Make your own play dough neuron by mixing 1 cup of flour, ½ cup of salt, and 2 teaspoons of cream of tartar. Once mixed thoroughly, add 1 cup of cold water with food coloring of choice and 1 tablespoon of salad oil. Thicken the mixture over low heat on the stove. Let it cool enough not to burn your skin and then knead the dough. Mold your dough into a neuron.
3. Pipe cleaner neuron. Obtain five different-colored pipe cleaners. Each pipe cleaner will represent a different part of the neuron. The parts are myelin sheath, dendrites, cell body, axon, and synaptic terminal.
   A. Roll one of the pipe cleaners into a ball (cell body).
   B. Take another pipe cleaner and push it into the ball and secure one end in the ball (axon).
   C. Push other pipe cleaners into the ball. These pipe cleaners should be shorter (dendrites).
   D. For the myelin sheath, wrap a pipe cleaner around the axon.
   E. Lastly, wrap a pipe cleaner around the end to represent the synaptic terminal.

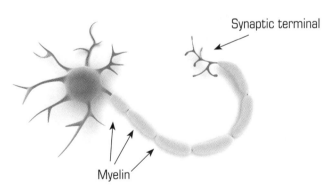

Word Scramble: Major Regions of the Brain

Please unscramble the words below that are included in the major regions of the brain:

1. oursCp llmCsuoa _____

2. uTamalhs _____

3. hHtmpsauolay _____

4. nBiar etSm _____

5. nosP _____

6. Mdauell Ontgoblaa _____

7. Putriaity _____

8. Siapln doCr _____

9. bleeelmruC _____

10. rbiidanM _____

11. rraeCebl xrteCo _____

Anatomy of the Central Nervous System

Color and label the parts of the brain.

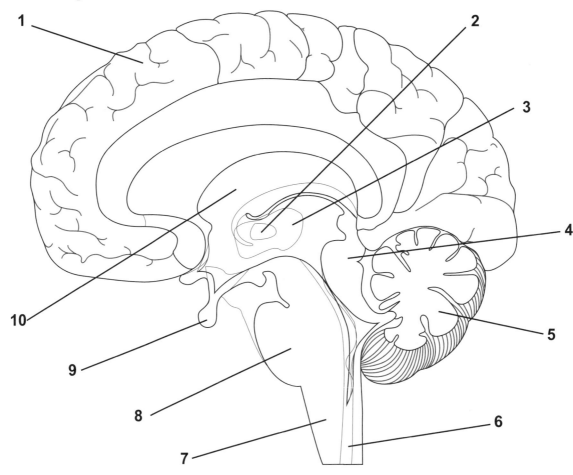

1. _____
2. _____
3. _____
4. _____
5. _____
6. _____
7. _____
8. _____
9. _____
10. _____

The Electrifying Nervous System | After Pages 26–29 | Day 18 | Worksheet 10 | Name

The Cerebrum

Fill in the blanks with the following words:

cerebrum   cerebral hemispheres   gray matter   white matter   corpus callosum

The right and left _____ are located in the _____.

The _____ contains the processes represented by the homunculus.

The area in which the neurological tracts are located is the _____.

_____ is the region in which the right and left hemispheres are connected.

Match the cerebral lobe with its function.

_____ Temporal Lobe   1. Personality, judgment, abstract reasoning, social behavior, and movement

_____ Parietal Lobe   2. Conscious perception of touch, pressure, vibration, pain, taste, temperature, memory processing, and conscious and subconscious regulation of skeletal movement

_____ Frontal Lobe   3. Visual cortex

_____ Occipital Lobe   4. Auditory cortex, olfactory cortex, language comprehension

(Continued on next page.)

1. Explain the location of the white matter and gray matter and their significance to how the brain functions.

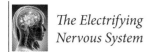

The Electrifying Nervous System | After Pages 26–29 | Day 18 | Worksheet 11 | Name

What Type of Brain Do You Have?

The right and left hemispheres of the brain have distinctly different functions. People are sometimes described as being right- or left-brained. Of course, both sides of our brains operate at the same time.

1. Based on the description of the right and left "brained" person on page 27, which type of person would you call yourself?

_____
_____
_____
_____
_____
_____
_____
_____

2. What potential problems or conflicts may be experienced between right- and left-brained people?

_____
_____
_____
_____
_____
_____
_____
_____
_____
_____

3. What does the Bible say about handling conflict? (Hints: Matthew 18:15–17; Matthew 5:9; Ephesians 4:26–27)

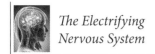

| The Electrifying Nervous System | After Pages 26–29 | Day 19 | Activity 9 | Name |

Jingle, Jangle

Make a product and slogan for something fictional to improve memory, sense of humor, or intelligence. It can be a circular advertisement, jingle, slogan, or commercial.

For example: Smart Chocolate

Brain Transplant

Oral Presentation: If brain transplants were possible in the case of injury, would you be for or against such a surgical procedure? Explain your rationale.

Head Injuries

Public service announcements are public announcements meant to help prevent or solve problems. Many are made by government or charitable agencies and may focus on a variety of issues, including health issues (don't smoke or do drugs, exercise more to be healthy) or solving environmental problems (don't litter, don't waste water, plant trees, recycle).

Oral Presentation: Make a 20-second public service announcement for one of the following topics: bike helmets, shaken baby syndrome, or drug abuse.

A Bird on the Wire

A carrier pigeon is a domesticated pigeon, trained to carry messages and then return to home. They were often used in the past to carry messages during war before the invention of radio and other forms of communication, and even used in races and to carry urgently needed medications.

Compare and contrast carrier pigeons with neurons.

The Electrifying Nervous System | After Pages 30-32 | Day 21 | Worksheet 12 | Name

Regions of the Brain

The cerebrum has four distinctive functional areas. These areas are composed of white matter.

Draw a picture demonstrating a function controlled by each of the lobes.

Lobes	Functional Picture
Frontal	
Parietal	

Temporal	
Occipital	

| | The Electrifying Nervous System | After Pages 30-32 | Day 21 | Worksheet 13 | Name |

Action and Control

Each of the actions below is controlled by a particular region of the brain. Place a check mark in the corresponding boxes that control the action.

Controls

Action	Frontal Lobe	Occipital Lobe	Parietal Lobe	Temporal Lobe
Language Comprehension				
Memory				
Visual Cortex				
Personality, Judgment				
Social Behavior				
Coordinates Movements				

The Frontal Lobe

Research the life of a railroad worker in the 1800s. Write a diary entry from the perspective of one of Phineas Gage's workers who were on the scene that infamous day on September 13, 1848. Include also how you came to work on the railroads and your relationship with Phineas.

(PD-Art)

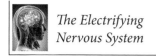

| The Electrifying Nervous System | After Pages 30-32 | Day 22 | Activity 13 | Name |

Write It Out

The many activities we perform with our bodies are done "automatically." Many tasks we do without thinking. In this scenario, you encounter a minion from the planet Zyataxon. He is confused about the customs of the planet Earth. You are dining on a peanut butter and jelly sandwich. He wants to know how to make one. The only form of communication he has been tutored to understand is the written English language. Write out the steps on how he can construct his own sandwich. Add as much detail as possible, including any neurological activities that occur during this process.

Step 1:

Step 2:

Step 3:

Step 4:

Act It Out

Imagine the same scenario as the previous page, but the Zyataxonian minion cannot understand spoken or written language. Act out how to make a peanut butter and jelly sandwich so that he can understand.

The Life of Dr. Penfield

Write a one-page biography on Dr. Wilder Penfield. Use this worksheet to gather your facts.

A biography is an account of a person's life. It highlights important events or experiences that made the person's life significant.

1. Where was he born?

   _____

2. When was he born?

   _____

3. Where did he grow up?

   _____

4. Where did he receive his education?

   _____

5. Did he ever get married?

   _____

6. Did he have any children?

   _____

7. Was he a man of faith?

   _____

8. Who inspired him to go into the field that made him famous?

   _____

9. What is he famous for?

   _____

10. What human qualities were most influential in shaping the way this person lived and influenced his times?

    _____

Dr. Wilder Penfield

11. On whom was his most difficult patient to operate?

Use the answer to the following question as your conclusion:

12. What are two or three important lessons you or any other young person might learn from the way this person lived?

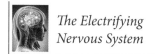

The Electrifying Nervous System | After Pages 33–35 | Day 23 | Worksheet 16 | Name

The Homunculus

1. What is the meaning and the origin of the word *homunculus*? What does the homunculus represent?

   _____
   _____
   _____
   _____
   _____
   _____
   _____

2. What is the difference between the motor homunculus and the sensory homunculus?

   _____
   _____
   _____
   _____
   _____
   _____
   _____
   _____
   _____
   _____

| The Electrifying Nervous System | After Pages 33–35 | Day 24 | Activity 15 | Name |

The "Little" Man in the Brain

Build your own homunculus. Draw a rough draft of your very own homunculus using the proportions seen in the illustrations of the text. Once completed, construct a model of your homunculus using your picture as a guide.

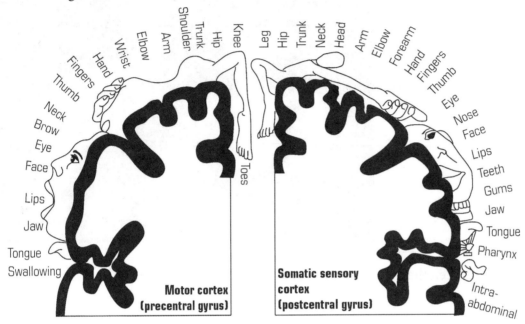

Name the major regions of the brain:

1. _____

2. _____

3. _____

4. _____

5. _____

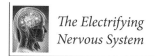

The Electrifying Nervous System | After Pages 33–35 | Day 25 | Activity 16 | Name

Broca's and Wernicke's Areas

You are a speech pathologist. A speech pathologist is one who specializes in communication disorders and helps rehabilitate patients with swallowing disorders. You are interviewing two new patients who are being admitted to the rehabilitation ward. One patient has suffered a stroke in Broca's area and the other suffered a stroke in Wernicke's area. Write five questions that you will ask each patient, and predict how they will act and respond to your questions.

Question 1:

Question 2:

Question 3:

Question 4:

Question 5:

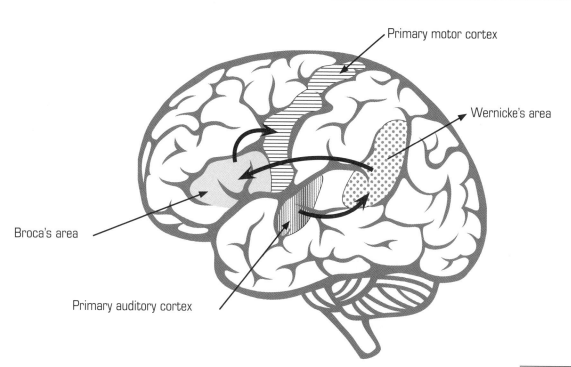

Penelope's Lost Puppy

You will need two crayons that are different colors, like red and blue, green and orange, or any color combination you would like.

In the following short story, underline in one color when Broca's area is at work.

Underline in a second color when Wernicke's area is at work.

Searching for Pretty

Penelope's little poodle named Pretty went missing one day. Penelope began searching the neighborhood and talking to any of the neighbors that might have seen little Pretty that day.

The first neighbor she met was Miss Margaret who lived next door and liked to raise flowers. "Miss Margaret," Penelope said. "Have you seen my little Pretty today?"

Miss Margaret looked up from behind the rose bushes, thought for a moment, and said, "I think I saw Pretty prancing down in Mr. Pulman's yard."

"Hmmm," said Penelope, "Mr. Pulman's yard? Okay, I will check with him — thank you, Miss Margaret."

Penelope skipped down the sidewalk a short distance and happened to meet Mr. Pulman on the way to get his newspaper at the end of the driveway. "Mr. Pulman," Penelope asked, "Have you seen my little poodle named Pretty?"

Mr. Pulman laughed and said, "I think Pretty is playing in the back yard with our little puppy named Pauley. They are jumping in and out of Pauley's kiddie pool."

Penelope jumped for joy and raced around to the back yard to get Pretty and take her home. She made sure to tell her parents about Pretty's adventure that evening at supper.

| The Electrifying Nervous System | After Pages 36–39 | Day 27 | Worksheet 17 | Name |

The Brain and Growth

The skull is like a "jigsaw puzzle" collection of bones. The **sphenoid bone** houses the **pituitary gland** in the skull.

The pituitary is a gland that lies within the brain. It is a vital gland that helps to regulate growth. Pictured to the left and below are two of the shortest and tallest recorded men in the world.

These two men met at the 10th Guiness World Records Day. The late Bahadur Dangi from Nepal measured at 1 foot and 9.5 inches and weighed 32 pounds. Sultan Kosen from Turkey, below, towers at 8 feet and 3 inches and weighs 302 pounds.

CC BY-SA 3.0

Write a one-page report about the pituitary.

Include the following details:

- Location of the pituitary
- What are some of the hormones produced in the pituitary?
- What are two functions of the pituitary?
- List the disorder that can result from a malfunctioning pituitary.
- What common things would be exceptionally challenging for each man?

CC BY-SA 2.0

Drug Effects

Research the effects of drugs on the brain: cocaine, crystal meth, alcohol, nicotine, and marijuana. Why are drugs so addictive, and where in the brain do they have the most impact?

NOTE: This activity will require supervised use of a computer and the Internet. Or you can get related books from the library or local health offices to help complete the work. There are a number of resources online that can help with this activity.

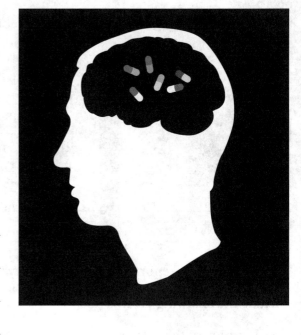

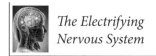

| The Electrifying Nervous System | After Pages 36–39 | Day 28 | Activity 19 | Name |

Brainiac

Fill in the blanks below.

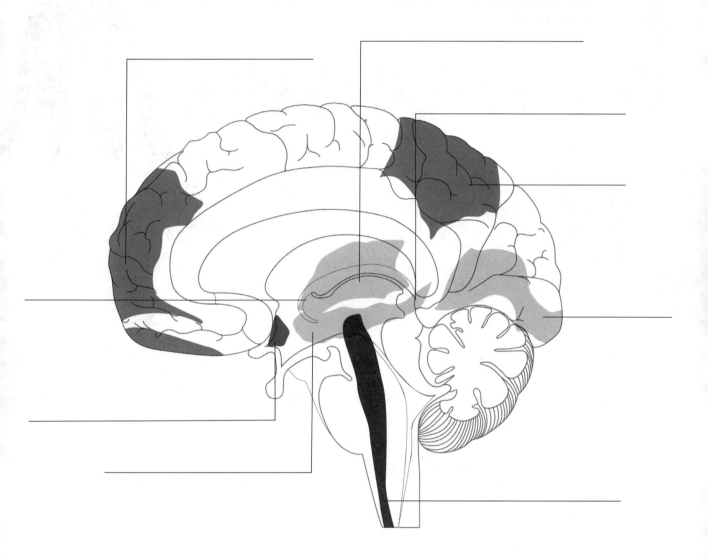

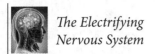

Training Your Cerebellum

In sports, athletes train their bodies with precise movements in order to navigate their sports. This training takes hours of work. It is through this training that the cerebellum helps maintain the muscle movements so that the movements can be executed the same way each time.

As a finely-tuned athlete, you will go into a bit of training. Pat your head and rub your belly, then try switching hands and repeat. Initially, this may be performed with much difficulty, but with practice you will be able to execute this flawlessly.

Fearfully and Wonderfully Made

For you created my inmost being;
you knit me together in my mother's womb.
I praise you because I am fearfully and wonderfully made;
your works are wonderful,
I know that full well.
My frame was not hidden from you
when I was made in the secret place,
when I was woven together in the depths of the earth.
Your eyes saw my unformed body;
all the days ordained for me were written in your book
before one of them came to be.
Psalm 139:13–16

God has cleverly crafted each of us. He has placed His divine fingerprints on each of us. Each system in our body undergoes unique changes during development. Next to each picture in the following sequence, write an element that develops at that particular time in gestation.

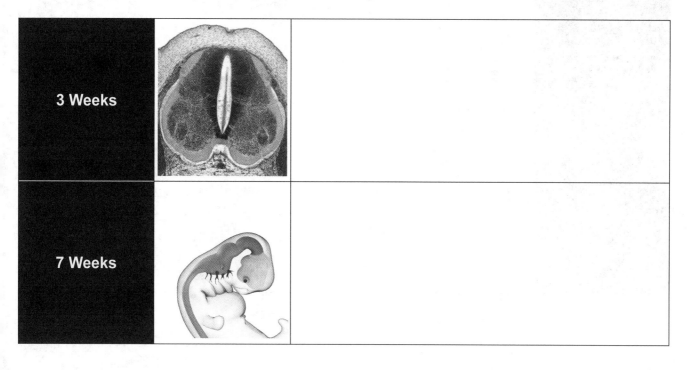

11 Weeks		
At Birth		
Infant Years		
Teenage Years		
Adult Brain		

 *The Electrifying Nervous System* | After Pages 40–43 | Day 32 | Activity 21 | Name

Brain Disorders

Oral Presentation: Compare and contrast Parkinson's and Alzheimer's diseases.

Here is a quick explanation of both conditions; you will need to do additional research to complete the activity with more detail.

Parkinson's disease is a condition where neurons in the brain start to break down and fail, leading to the loss of a chemical that helps carry messages from the brain. It is a progressive disorder that may begin with a slight shaking in your hand, and will eventually lead to being unable to move or walk or speak. It cannot be cured, but medicines are being developed that help to lessen the effects of the condition.

Alzheimer's disease is a brain disorder that destroys memory and the ability to think, and it eventually leads to being unable to perform even the very simplest of tasks that are common to daily life. It impacts a person's ability to reason and even speak. Three effects of the disease in the brain are loss of connections between neurons, abnormalities, and tangled bundles of fibers.

Dough Brain Model

Materials:

Gray-colored paint (water-soluble)
Newspaper
Masking tape
Paint brush
School or wood glue
Soft white bread

1. Take several sheets of newspaper and crumple them into an oblong ball approximately the size of a grapefruit.
2. Pack the newspaper tight, then wrap masking tape around the ball to help it keep its form.
3. Put the glue in a paper cup and mix the gray-colored paint. Using the paint brush, coat the form with glue.
4. Remove the crust from the bread. Squeeze and roll the soft white bread in a cylindrical tube. Glue the bread on the form in various configurations to signify the gyri of the brain, and coat with the gray glue mixture.

After making your project, point out main structures and areas of the brain to your instructor or label these areas to demonstrate your understanding.

Don't Be an Egghead!

In your amazing container, the skull, your brain floats in a clear substance called cerebrospinal fluid. One of its functions is to protect your brain from any sudden impact. This experiment will be a demonstration on how this works.

Materials:

> 2 raw eggs
> Permanent markers (waterproof)
> Plastic container with a lid
> Water

1. Draw a face on your eggs. The inside of the egg represents the brain, and the actual eggshell will represent the pia mater. The pia mater is the inner layer of the meninges, which covers the brain.

2. Place your "egg head" in the plastic container with a lid. The container should be slightly bigger than your egg. This container will represent the skull. Now shake the container vigorously. Take the egg out and examine the results.

3. Next, repeat the same experiment with the other egg, but this time fill the container with water and put the egg in the water of the container and seal it with the lid. Now shake the container again. Take this egg out and examine the results.

Did you see any differences in the eggs? What does the water represent?

You can design your own experiment by placing the egg in the container with and without water, and dropping the egg container from varying heights. Record your findings. Make predictions prior to your experiments. In addition, you could try different fluids or materials in place of the water — for example, Styrofoam™, cooking oil, sand, etc.

| The Electrifying Nervous System | After Pages 44–45 | Day 35 | Activity 24 | Name |

Holes in the Head

Our skulls are filled with holes. These holes, or foramen, serve an important purpose of allowing nerves to travel through the skull like a tunnel. The nerves travel through and innervate (form a neurological connection that joins structures) the brain. Examples of such foramen are the *foramen sphenoid*, *foramen magnum*, and *foramen ovala*.

1. Research the location of one of these foramen, find the name of a key nerve that runs through it, and identify what it connects (innervates).

2. Explain the effect a broken skull at that point would have on the nerves and the muscles that are innervated by those specific nerves.

_____
_____
_____
_____
_____
_____
_____
_____
_____
_____
_____

Experiment with the Blood-Brain Barrier

CAUTION: Using food coloring or dyes can stain hands and clothing. Protect surfaces as well. Be very careful and ask an adult for help when using it.

Materials:

Lab Instruments	Substances	Other
Test-tube rack	Clear cooking oil	Marking pen
Safety goggles	Sesame or motor oil	Masking tape
Six test tubes with stoppers	Water	Paper towels
Three eye droppers	Red and blue food coloring	Colored pencils (red and blue)
Funnel (optional)	Alcohol	Lab report form
		Cup for mixing water with food coloring

Introduction:

The Blood-Brain Barrier (BBB) helps protect the brain from many harmful things, such as chemicals, drugs, and bacteria. We can visualize the BBB as a sentry guard standing at the gate to protect the delicate neuron cells and allowing only selected nutrients to pass through the gate. In this experiment, you will explore this process. The clear cooking oil will represent the BBB, and the sesame or motor oil will represent a fat-soluble substance. The water with a little red food coloring will represent blood. Finally, the blue food coloring will represent a water-soluble substance.

Background Work:

Research the following terms: water-soluble and fat-soluble substances. Clearly write out these definitions.

_____

_____

_____

Procedure:

1. Label four test tubes as 1, 2, 3, and 4. Feel free to use masking tape to assist with the labeling.
2. With the aid of a small funnel (optional), fill each test tube one-fourth full with clear cooking oil. In a cup, mix a small amount of water with a few drops of red food coloring. Place red water into each test tube, filling them to the halfway mark.

Complete the setup as explained below:

Test Tube 1	Completed. Place a stopper on top.
Test Tube 2	Add 8–10 drops of blue food coloring and put a stopper on top.
Test Tube 3	Add 8–10 drops of sesame or motor oil and put a stopper on top.
Test Tube 4	Add 8–10 drops of sesame or motor oil and 8–10 drops of blue food coloring. Place a stopper on top.

3. Record your observations.

_____
_____
_____

4. Predict what would happen if you were to shake tube No. 4 for 15 to 20 seconds. Then, shake the tube. Do your observations match your predictions?

_____
_____
_____

5. Label the remaining two test tubes No. 5 and No. 6. Complete the setup as explained below:

Test Tube 5	Fill the test tube one-fourth full with cooking oil, and then fill to the halfway mark with alcohol.
Test Tube 6	Fill the test tube one-fourth full with red colored water, and then fill to halfway mark with alcohol.

6. Observe the tubes for a minute or two. What happens? Record your observations.

_____
_____
_____
_____

7. Complete the lab form.

_____
_____
_____
_____

| *The Electrifying Nervous System* | After Page 46 | Day 37 | Worksheet 19 | Name |

The Blood-Brain Barrier Maze

This feature of your body helps to protect your brain and spinal cord. It works like a security system to keep bad things out but still let good things through. See if you can guide the policeman out of the maze and think about how hard your blood-brain barrier works to protect you!

Circle the words in this word cloud that are things that can harm the blood-brain barrier:

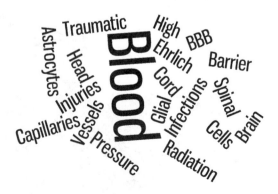

Worksheet 19, Day 37 // 91

Page left intentionally blank.

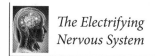

The Electrifying Nervous System — After Pages 47–49 — Day 40 — Worksheet 20 — Name _____

The Backbone

Use the following words to fill in the blanks:

Vertebral   Column   Lumbar   Sacral   Virus

Dermatomes   Shingles   Nerve root   Spinal ganglion   Thoracic

Cold sore   Dormant   Sensory   Cervical

1. _____ is related to the chicken pox virus.

2. If a virus "goes to sleep" it is said to be _____.

3. List the four regions of the vertebral column:

    a. _____

    b. _____

    c. _____

    d. _____

4. Tightly packed groups of nerve cells at the site in which the nerve exits the spinal cord is called _____.

5. Regions of the skin are connected to a pair of separate nerve roots called _____.

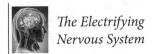

Stem Cells

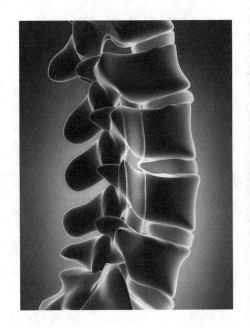

Oral Presentation: Stem cells are special cells in your body that can turn into different kinds of more specialized cells and then divide to produce more stem cells. They have been used for research and treatment purposes, including some that are related to muscle or nerve injuries.

Prepare a speech (pro or con) defending why using stem cells in spinal cord injuries is an acceptable or not acceptable proposition.

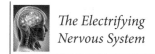

| The Electrifying Nervous System | After Pages 47–49 | Day 41 | Activity 27 | Name |

Spinal Cord Dissection Lab

CAUTION: This activity requries adult supervision! A scalpel or knife is not a toy. When not making the needed cuts to the specimen, the scalpel or knife should be carefully laid down where it will not fall off or be able to be reached by younger students.

Peer into the inner workings of a spinal cord.

Materials:

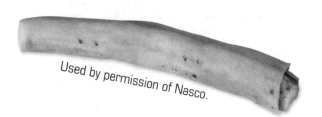

Used by permission of Nasco.

- Cow spinal cord
- Scalpel
- Dissecting tray
- Probe
- Gloves
- Magnifying glass

Instructions:

1. Lay the specimen on a tray. The first step in any good dissection is to use your power of observation. Observe any external structures on the spinal specimen segment. Draw your observations below.

2. With a scalpel or knife, cut the spinal cord in half horizontally. Observe the coloration of the central aspect and other aspect of the cord. Are there any differences? What do you think would cause these differences? Draw your observations.

3. With a probe, tease a section of specimen apart. What consistency does it have? Does it appear something like "string cheese"? Why do you think this is?

| The Electrifying Nervous System | After Pages 47–49 | Day 41 | Activity 28 | Name |

Spinal Column Model

The vertebral column is composed of numerous joints that allow the body to bend in various directions. The spinal cord is housed in this tall column for protection.

Materials:

5–6 empty thread spools
Yarn (any color)
Hole punch
Masking tape
Foam or cardboard disks cut to the size of the spools
Drinking straw

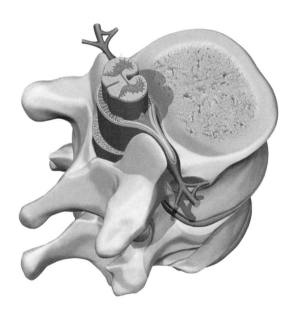

Directions:
1. Take the foam or cardboard disks and punch a hole in the center of each.
2. Cut yarn into 4- to 5-inch segments. Cut 2 strands for each spool.
3. Take 2 strands and thread into each spool.
4. With yarn remaining in each spool, push the straw through each spool and disks alternating between each.
5. Secure the top and bottom of your spinal column model with masking tape.

Questions:

1. What does each of the items in your spinal column model represent?

   Yarn:_____

   Disks:_____

   Spools:_____

   Straw:_____

2. What purpose do the disks serve in the vertebral column?

   _____
   _____
   _____
   _____
   _____
   _____
   _____
   _____
   _____
   _____
   _____
   _____
   _____

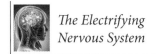

The Electrifying Nervous System | After Pages 47–49 | Day 42 | Activity 29 | Name

If You Couldn't

Imagine that you have injured some part of your backbone and spinal cord and you now struggle to accomplish the tasks and hobbies you may enjoy. In the space below, draw an invention that would help you to accomplish something you like to do, even if you were in a wheelchair or had limited use of your arms or legs.

All Circuits Firing

Fill in the Blank

1. Jogging _____ a reflex (*is, is not*)
2. Most of the body's reflexes are _____. (*fast, slow*)
3. You are _____ to control your reflexes. (*able, not able*)
4. A reflex happens _____. (*always in the same way, changes depending on your mood*)
5. Reflexes are _____. (*conscious, unconscious*)
6. When you step on a nail you demonstrate the _____ reflex. (*overdraw, withdraw*)
7. When a baby's foot is stroked, the big toe turns _____ (*upward, downward*)
8. A baby turning its head in the direction of its cheek being stroked is called the _____ reflex. (*rooting, uprooting*)

Word Scramble

Unscramble the words below.

1. xoan _____
2. ddtenrsei _____
3. rennuos _____
4. ltpcociia _____
5. mbecerur _____
6. uuuhmnolsc _____
7. gsaarotli _____
8. sspynae _____
9. mnnieegs _____

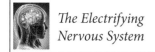

The Electrifying Nervous System | After Pages 50–53 | Day 44 | Activity 30 | Name

Reflexes — Reaction Tester

Materials: Ruler and a partner

Instructions:

Ask your partner to hold a ruler in the vertical position. You will place your open hand ready to catch it, down below the ruler. When your partner drops the ruler, you will attempt to pinch your fingers to catch the falling ruler as quickly as possible. Record the position of your fingers on the ruler. Repeat this experiment several times, and record the results. Did you note any improvement in your reaction time when catching the ruler? Explain the results you find.

_____

_____

_____

_____

_____

_____

_____

_____

_____

_____

_____

_____

_____

It Is All in the Timing

Doctors assess reflexes to determine whether nerves are working properly. A reflex is an involuntary (not conscious) quick muscle movement in response to a stimulus. Through reflexes, our bodies can protect themselves. There are many areas on the body in which reflexes can be elicited. Try a few on a willing participant. It works best if your examinee is relaxed.

Materials: Reflex hammer and a willing subject

1. Patella Reflex: Have your subject sit on the edge of a table with his or her legs hanging freely. His or her feet should not touch the floor. Find the patella tendon. (This tendon connects the quadriceps muscle to the leg bone; the patella lies within this tendon.) It can be located about one finger width down from the kneecap. Gently tap this region with a reflex hammer to elicit a response. Your subject needs to be completely relaxed and feet freely hanging off the ground on an elevated stool. If the subject is tense, the reflex cannot be elicited. If your subject is having difficulty relaxing, ask him or her to put his or her hands together. With the subject's fingers interlocked, grasp his or her hands and pull while he or she is pulling on his or her own hands. This task will distract your subject and may allow you to elicit the reflex.

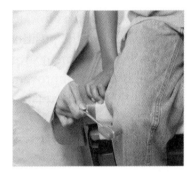

2. Biceps Reflex: Have your subject sit in a comfortably relaxed position. Support your subject's arm by cradling it in one of your arms as illustrated below. First, isolate the biceps tendon on the inner (medial aspect) part of the arm in the cubital fossa (the top side and fold of the arm at elbow). Place your thumb over the area that feels like rubber in this region. Make sure your subject's sleeve is rolled up to allow you to observe the muscle and watch the lower arm for movement. A normal response will cause the biceps to contract, drawing the lower arm slightly upward.

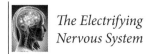

The Electrifying Nervous System — After Pages 50–53 — Day 46 — Activity 32 — Name

Light as a Feather

This experiment can be done with a partner or by yourself. Ask your partner to stand behind you. With your arms straight down, try to lift your arms as your partner holds your arms down. Do this for 30 to 45 seconds as hard as you can. If you do not have a partner, stand in a doorway and attempt to lift your arms while pressing them into the frame for the same amount of time.

Now, release the pressure and allow your arms to hang. What happens next? Why do you think this happens?

The Scary Stuff Challenge

Choose one of the following activities:

- List 5 things that you are scared of or find scary, and then create a song about not being afraid of them.

- Create a scary story you can tell your family or class using the following details in the story — a cloudy day, a plate of spaghetti, a fruit bat, a pair of purple sneakers, and a old playhouse.

- Draw a scary cartoon or drawing.

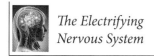

The Electrifying Nervous System | After Pages 54–55 | Day 47 | Worksheet 22 | Name

The Hypothalamus

Short Answer Questions

1. What does your hypothalamus do?

2. In what part of the brain is the hypothalamus located?

3. How do our bodies fight bacterial or viral infections?

4. What is the substance that your body releases to tell your hypothalamus to warm up your body's temperature?

5. What is the name for the roof of the mouth? And why do you get a "brain freeze"?

Bonus Fun

See if you can say Sphenopalatine ganglioneuralgia five times fast!

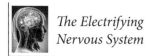

| The Electrifying Nervous System | After Pages 54–55 | Day 48 | Activity 34 | Name |

Memory

Materials: 10 small random objects, large towel

Instructions:

1. Try this game with your family or friends. Locate various random small objects around your home (pen, small bouncy ball, forks, hairpin, etc.) and place the objects on the table.
2. Take the towel and cover the objects.
3. Allow a family member or friend to see the uncovered objects on the table for approximately five to ten seconds.
4. Re-cover the objects. Give the person 30 seconds to recall the items by memory.
5. Do this again. Take one object away without the person seeing. Now, uncover the objects and see if he or she can remember the one missing item.
6. Which challenge is harder? If you give clues to your family member or friend, is he or she able to recall the objects? What helpful strategies did he or she use to recall the objects?

Take a Picture

You will need a partner.

Obtain a picture from a magazine or book. Describe the picture to a partner without telling him or her what it is. Have your partner draw the picture based on your description. When you are done, compare the picture to your partner's rendition.

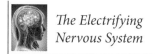

The Electrifying Nervous System | After Pages 56–58 | Day 50 | Worksheet 23 | Name

A Personal Sleep Study

1. What time do you normally go to sleep during the week? On the weekend?

   _____
   _____
   _____
   _____

2. What do you normally do before you go to sleep (i.e., brush your teeth, take a bath, etc.)?

   _____
   _____
   _____
   _____

3. On average, how many hours a night do you sleep?

   _____
   _____
   _____
   _____

4. Are there certain pillows, blankets, or toys you need to have with you while you sleep?

   _____
   _____
   _____
   _____

5. Do you sleep in the dark, or do you use a nightlight?

   _____
   _____
   _____
   _____

6. Do you ever go to sleep with the TV, radio, or other source of music playing?

_____
_____
_____
_____

7. Do you ever take naps during the day?

_____
_____
_____
_____

8. Do you have a pet that sleeps on the bed? Does this ever wake you up?

_____
_____
_____
_____

9. Do you dream when you sleep? Guess about how many times a week you may dream.

_____
_____
_____
_____

10. Does someone in your family go to bed later than you? If so, does it wake you when he or she goes to bed?

_____
_____
_____
_____

Fun with Limericks

A limerick is a funny poem. Limericks have not always been around. The first limericks were published by Edward Lear (1812–1888) in a book entitled *A Book of Nonsense*. He was an English humorist and painter. A true limerick has the following basic ingredients. To qualify as a limerick it must:

- Have five lines
- The first, second, and fifth lines all rhyme with each other
- The third and fourth lines rhyme with each other
- The lines have an unique rhythm
- They are usually humorous

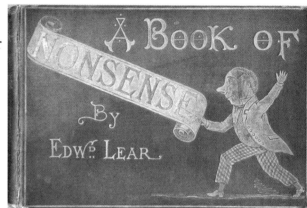
(PD-US)

Rhyming Structure and Rhythm

A limerick has a particular rhyming structure known as "AABBA."

The limerick also processes an anapaestic rhythm. Anapaestic has two short syllables followed by a long one. The rhythmic beat sounds a bit like the following:

    da DUM da da DUM da da DUM
    da DUM da da DUM da da DUM
    da DUM da da DUM
    da DUM da da DUM
    da DUM da da DUM da da DUM

Here is an example:

    There was a Young Lady whose chin
    Resembled the point of a pin:
    So she had it made sharp,
    And purchased a harp,
    And played several tunes with her chin.
                    — Edward Lear

Now do your own limerick on the next page!

There once was a

_____8 syllables

_____8 syllables

_____5 Syllables

_____5 Syllables

_____8 syllables

 *The Electrifying Nervous System* | After Pages 59–61 | Day 52 | Activity 37 | Name

Sleep

Using our observation skills, we learn a great deal about our world. Next time your pet or a person in your family falls asleep, quietly observe them. Can you determine whether they are dreaming? Can you see any movement behind his or her closed eyelids?

Look for clues:

- Body movement
- Eye movement
- Sounds
- Facial expressions

| | After Pages 59–61 | Day 53 | Activity 38 | Name |

Sleep Simulator

There are five stages for sleep. See if you can recreate them, but don't fall asleep! Lay down on the couch and see if you can close your eyes and imagine going through the five stages. Write in the final column if you were able to feel similar to the description of the stages.

Stage 1	When you are barely asleep, almost like daydreaming.	Lay down and close your eyes; try to relax for a few moments.
Stage 2	When we are slowly becoming unaware of the sights and sounds around us, more relaxed, breathing rate is normal and regular, and our body temperature starts to drop a little.	Did you experience that moment when you were completely unaware of things happening around you?
Stage 3 Stage 4	When we are breathing slower, very relaxed muscles, deep and restful sleep, our energy reserves are being re-energized, the muscles are getting more blood, and tissue repair is underway.	Imagine your body busily repairing itself as you rest. Did you become so relaxed you didn't want to move?
Stage 5	Our closed eyes are moving back and forth, our muscles are not moving, we are dreaming, and the cycle continues around every 90 minutes after we fall asleep.	Be very still, then look from side to side with your eyelids closed. Try it with one eye as you look in the mirror. It will show you what REM looks like.

Brain Food

Hint: If it has a seed, think fruit!

My Plate was adopted by the U.S. government to help promote healthy eating. The Plate illustrates the five basic food groups that compose a healthy diet. Identify the type of food in the choices below — for example, would a glass of milk be a protein or a vegetable? Write the correct answer in the space below the images.

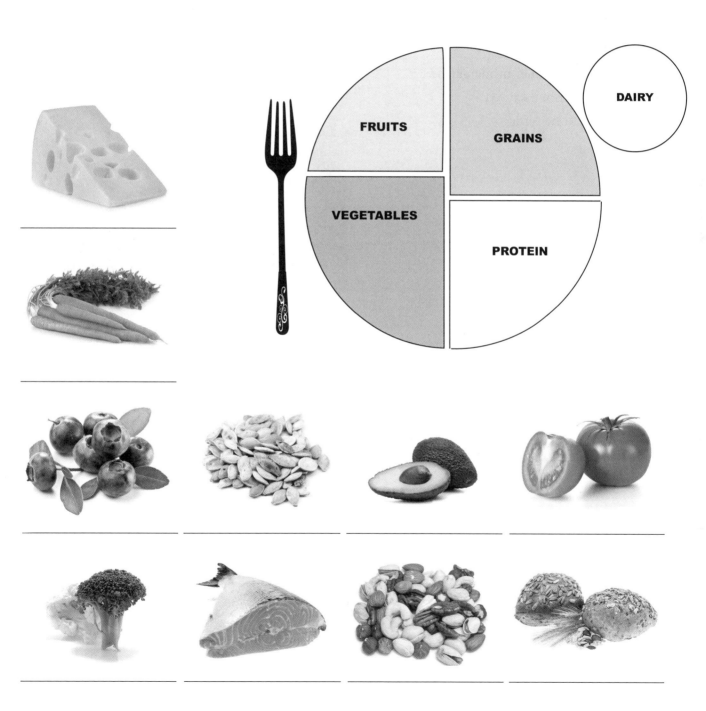

Coconut Lame Brains

Yummy, yummy for your tummy. This sweet treat will send a sugar rush right to your brain.

Ingredients:

- 3½ cups flaked coconut
- ¼ cup light cream
- 2 cups confectioners' sugar
- ¼ cup softened butter or margarine
- 1 teaspoon almond extract
- ¾ cup grenadine syrup

Materials:

- Mixing bowl
- Cookie sheet
- Spatula
- Tablespoon
- Waxed paper

Using the spatula, mix all the ingredients in a large mixing bowl EXCEPT for the grenadine syrup. A rubber spatula may work best. Next, line the cookie baking sheet with wax paper. This will allow for items not to stick to the baking sheet.

Scoop out the mixture with a tablespoon and press the mixture in a ball and shape into a brain shape. Ideally, the brain should be the size of an egg, just flattened on one side. Once your brains are formed, chill in the refrigerator until firm. This will take an hour or so.

Once chilled, drizzle with the grenadine syrup. Then serve your coconut lame brains.

Cross Over

Find the words hidden in the seek-and-find puzzle from the word bank below.

```
C Y U A X P C B A A C E R E B E L L U M L J H G E C A F C B
A X O N H Y O D C B F J E G F K H I J G M N S O R P Q R L D
D Z A B W C R N H B I D P Y R O G E N S O P T S R Q R O H I
J K E L V F P G S H I N G L E S L V K L M U N T O P O N Y E
H O M O N C U L U S U M A L A H T L A R O P M E T D X T P N
T L E C A B S E T I R D N E D D G Y R U S U T W B M Y A O C
S I N C U P C S E R E H P S I M E H L A R B E R E C C L T E
R G I I Z A A F I S S U R E S E D R W C E V A N B A S Z H P
P O N P S I L R A Z M A F E Z D D Y M X F I I D G H F S A H
O D G I E L L D I C N B B Q P U O N G A N N M L K J I N L A
Q E E T L G O F Y E L G R H C L S T R B T T I P I C C O A L
Y N S A C O S W H I T E V C B L D U A W X O G H Y Z I R M O
M D X L I R U A X B H A G I F A E R Y F E D M C G B A U U N
U O P Q R T M R W K I J L N E U R O G L I A J E K L J E S M
R G O V T S B S N I L E Y M M I C R O G L I A H S O I N J N
B L W V N A R B O R V I T A E P E N D Y M A L C E L L S P Q
E I V U E T O S T R U Q P R E S P A N Y S X W L V U T S K R
R A L O V K C I L J W E R N I C K E F E O N D M C Y K B Z A
E S N N O L A H P N E C S E M P I T U I T A R Y G L A N D A
C M R Q P N H M M E T S Y S S U O V R E N C I M O N O T U A
```

arbor vitae	astroglia	autonomic nervous system	axon
Broca	cerebellum	blood-brain barrier	cerebral hemispheres
cerebrum	corpus callosum	dendrites	dermatomes
diencephalon	ependymal cells	fissures	frontal
gray	gyrus	homunculus	hypothalamus
medulla	meninges	mescenphalon	microglia
myelin	neuroglia	neurons	occipital
oligodendoglia	parietal	pituitary gland	pons
pyrogens	shingles	synapse	temporal
thalamus	ventricles	white	Wernicke

| The Electrifying Nervous System | After Pages 69–71 | Day 59 | Worksheet 26 | Name |

Do You Know?

We have taken quite a journey through *The Electrifying Nervous System*. Write down five facts that you learned during your study of the nervous system.

1. _____
_____
_____
_____

2. _____
_____
_____
_____

3. _____
_____
_____
_____

4. _____
_____
_____
_____

5. _____
_____
_____
_____

The Electrifying Nervous System | After Pages 69–71 | Day 60 | Activity 40 | Name

Brain Strain

Our brains are amazing! They help us solve puzzles, remember things, see matches, and so much more! Here are some trick questions to exercise your brain.

Example:

Q. What will you find at the end of every rainbow?

A. The letter "w."

1. What two things can you never eat for breakfast?

2. What word is spelled incorrectly in every single dictionary?

3. What goes up but never comes down?

4. Which weighs more, a pound of feathers or a pound of bricks?

5. What 5-letter word becomes shorter when you add two letters to it?

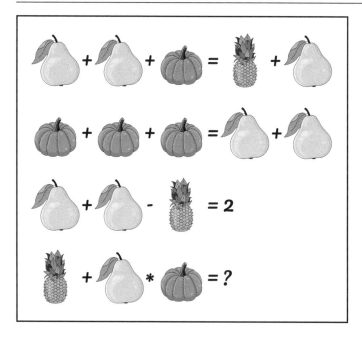

Extra Challenge

In this exercise, each type of fruit represents a number. With the hints given, can you solve the last math problem? (Hint: Don't forget about the order of operations!)

Trick questions taken from © 2020 The Thought & Expression Company, LLC. All rights reserved. Powered by WordPress.com VIP. https://thoughtcatalog.com/january-nelson/2018/03/65-riddles-for-kids-guaranteed-to-stump-you-too/

Shrunken Apple Head

Ingredients:

- Apple
- Butter knife
- Cup
- Newspaper
- Pencil
- Rice, uncooked
- Thumb tacks
- Vegetable peeler

Instructions:

1. Utilizing the apple peeler, remove the peel from the apple. Push the pencil halfway into the bottom of the peeled apple. This will serve as a holder for the apple.
2. Carve the facial features on the apple: nose, mouth, and two eye sockets with the knife. Be sure to get an adult to help.
3. Place a thumb tack into each of the eye sockets.
4. Place the grains of the rice into the mouth to represent teeth.
5. Wad up newspaper and press into the bottom of the cup. Take the apple impaled on the pencil and stick the pencil into the waded paper. Be careful not to allow the apple to rest on the lip of the cup. This will allow the apple to dry out.
6. Air dry the apple. This process will take at least 2 weeks. Presto, your shrunken apple head is ready for display.

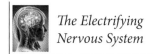 | *The Electrifying Nervous System* | After Pages 69–71 | Day 62 | Activity 42 | Name

Special Report!

I Am Wonderfully Made Because.....

You will be creating a short essay focused on one amazing fact that shows how incredibly well-designed the human body is. This is the first part of the exercise, helping you to choose and organize your thoughts for the second part of the exercise, actually writing the short essay.

Part 1: Getting organized

1. Choose one incredible design feature you have learned so far, and write it down. This will be the main focus of your essay.

   _____
   _____
   _____

2. What makes it so incredible or important to how your body works?

   _____
   _____
   _____

3. Draw a small example of this part of the body.

   _____
   _____
   _____

4. Are there specific names to different parts of this design feature? If so, write them.

   _____
   _____
   _____

| The Electrifying Nervous System | After Pages 69–71 | Day 63 | Activity 43 | Name |

Special Report!

I Am Wonderfully Made Because.....

Part 2: A short article

You are a reporter on assignment to write a short article on the human body. Write at least three short paragraphs about the designed feature of your choice, using the material you gathered in Part 1 of this activity.

_____
_____
_____
_____
_____
_____
_____
_____
_____
_____
_____
_____
_____
_____
_____
_____
_____
_____
_____
_____

Edible Oozy Goozy Brain Salad

SAFETY NOTE: Ask an adult to help when using the stove.

Ingredients

Aluminum foil
Colander
Cooking spray
Gray cake-decorating dye
Ice, crushed, ½ cup
Large plate
2 large pots
Measuring cup

Pasta sauce, 1 jar
Plastic wrap
Six ounces of spaghetti
Small bowl (the size of your head)
Spoon
Unflavored gelatin, 1 packet
Water

Instructions

1. Make several small wads out of the aluminum foil and place in your "brain" bowl.
2. With a large sheet of aluminum foil, line the bowl over the wads of foil. Press the foil firmly in the bowl, paying attention to shape the inner lining like a brain.
3. Spray the inner foil with the cooking spray. This will assist in removal of the "brain" once molded.
4. In the large pot, bring water to boil and break apart the spaghetti into several pieces. Place the broken spaghetti in the boiling water. Cook the spaghetti until it is tender.
5. In the other pot, begin preparation of the gelatin. Boil ½ cup of water and stir in the packet of gelatin. Once it is dissolved, turn the burner off. Stir in crushed ice.
6. When the gelatin begins to congeal, add the mix to the pasta in the other pot and stir it together. Place the gray food coloring in, one drop at a time, until the desired color is achieved.
7. Take the pasta and gelatin mixture and pour into the mold.
8. The mold will need to be covered in plastic wrap and refrigerated.
9. Once the oozy goozy brain has solidified, take the plastic wrap off and place a plate on top. Turn the mold over onto the plate. The brain should slide right out.
10. Spoon pasta sauce on and around the brain.
11. Bon Appetit!

This activity was more than just to make a treat! It's to help you recognize and see how the brain is structured.

Share your thoughts on the brain and what you have learned so far:

Gross Anatomy at the Brain Lab

SAFETY WARNING: Always have an adult supervising this or any activity in which you use the scalpel or a knife. When cutting with a scalpel or knife, make sure you always cut away from yourself or anyone else. **SET THE KNIFE DOWN CAREFULLY ON THE TABLE WHEN THE TASKS ARE COMPLETED — THIS IS NOT A TOY.**

Objectives:

Identify the structures of the brain and locate the functional areas.

Materials:

    Kitchen knife
    Dissection kit
    Preserved sheep brain with cranial nerves attached
    Dissection tray
    Disposable gloves

Instructions:

1. Examine the external structures of the brain.

2. Locate:
   A. Cerebral hemispheres — right and left
   B. Parietal lobe
   C. Frontal lobe
   D. Occipital lobe
   E. Temporal lobe
   F. Optic nerve
   G. Olfactory bulb
   H. Pituitary gland
   I. Medulla oblongata
   J. Spinal cord

Draw a picture of a specimen that identifies each of the above structures.

3. Place the brain with the ventral side up (top side up). The optic nerve and medulla oblongata are found on the posterior side (bottom side down). With the kitchen knife, slice between the right and left hemispheres, cutting completely through all structures. You will now have two halves. This cut will transect the brain tissue, corpus callosum, midbrain, cerebellum, and brain stem.

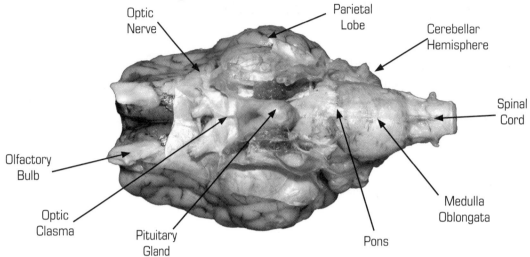

4. Place the brain in the tray with the newly cut side up. Now identity the structures in the picture above in your specimen.

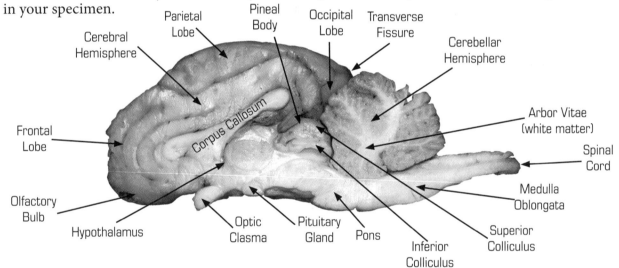

Complete the following chart (for Level 3 students), writing the function of each part:

Corpus Callosum	
Optic Chiasm	
Hypothalamus	
Pituitary Gland	
Medulla Oblongata	
Cerebellum	
Parietal Lobe	
Frontal Lobe	
Occipital Lobe	
Pineal Gland	
Thalamus	

126 // Elementary Anatomy, The Electrifying Nervous System

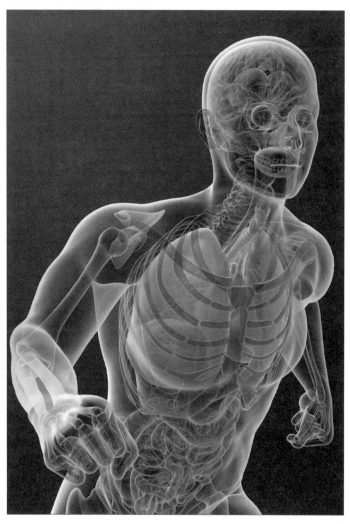

The Breathtaking Respiratory System

Educator Aids
for Use with
*Elementary Anatomy: Nervous,
Respiratory, and Circulatory Systems*

Page left intentionally blank.

The Breathtaking Respiratory System

RESPIRATORY SYSTEM OBJECTIVES

Successful completion of this module will enable the student to:

- Describe the primary function of the respiratory system.
- Describe the function of mucus.
- Name the location of the sinuses and their function.
- Name the respiratory structures and describe their functions.
- Describe the function of the nasal cavity and nostril hairs.
- Name the parts of the upper respiratory tract.
- Name the parts of the lower respiratory tract.
- Name the tissue comprising the rigid support of the larynx.
- Describe the function of the epiglottis and vocal cords.
- Describe the location and function of bronchi, bronchioles, and alveoli.
- Describe the function of cilia.
- Describe the function of the diaphragm.
- Describe how gas exchange occurs between the alveoli and pulmonary capillaries.
- Describe where the main respiratory control center is located.
- Describe how oxygen is transported throughout the body.
- Explain why carbon monoxide is poisonous.
- Describe some of the hazards of cigarette smoking.
- Describe the difference between an epidemic and pandemic.
- Describe one characteristic or fact for each of the following illnesses:

    Tuberculosis
    Cystic Fibrosis
    Allergic Rhinitis
    Asthma
    Laryngitis
    Polio

Page left intentionally blank.

The Breathtaking Respiratory System

SUPPLY LIST FOR THE ACTIVITIES

Activity 46: Respiratory Flash Cards
- [ ] Scissors
- [ ] Tape or glue stick

Activity 47: Snot O' Matic
- [ ] Borax Laundry Booster
- [ ] 1 liter or quart soda bottle, clean
- [ ] Measuring cup
- [ ] Green or neon food coloring*
- [ ] Tap water
- [ ] Tablespoon
- [ ] White glue

Activity 48: Wind Bag
- [ ] Large balloons (several)
- [ ] One 2-foot length of yarn or string
- [ ] Ruler
- [ ] Markers

Activity 50: Measuring Lung Capacity
- [ ] Large bowl of water
- [ ] Milk jug or 2 L pop bottle, marked
- [ ] Straws

Activity 51: Smoke in Your Lungs
- [ ] 1-L clear plastic bottle
- [ ] Cotton balls
- [ ] Clay or play dough
- [ ] Rubber tubing or straw
- [ ] Cigarette

Alternate List
- [ ] Spray bottle
- [ ] Water
- [ ] One color of dye*
- [ ] Straw

Activity 52: What's the Count?
- [ ] Stopwatch

Activity 53: The Voice of a Balloon
- [ ] Balloon

Activity 54: Breathe
- [ ] Paper lunch bag

Activity 55: Wind Bag
- [ ] Lung volume bag
- [ ] Mouth piece

Activity 56: What Is Your Girth?
- [ ] Tape measure

Activity 58: A Nose for Clues
- [ ] Bag of assorted hard candies
- [ ] Bandana or scarf

Activity 59: Grape-Like Clusters
- [ ] Balloons (small)
- [ ] Yarn (blue, red, yellow)
- [ ] String
- [ ] Rope
- [ ] White glue

Activity 61: Pulse and Respiratory Rates
- [ ] Stopwatch or timer
- [ ] Jump rope

Activity 62: The Air Around Us
- [ ] Sheet of graph paper
- [ ] Petroleum jelly
- [ ] Scissors
- [ ] Microscope slides (5)
- [ ] Pencil
- [ ] Tape

Activity 64: What Trees Want
- [ ] Test tubes (2)
- [ ] Test tube rack or holder
- [ ] Bromothymol blue solution
- [ ] Straw
- [ ] Stopwatch or timer
- [ ] Glass-marking pencil (China pencil)
- [ ] Jump rope

*Please be cautious with items that can stain surfaces and clothing.

Activity 66: Model of Lung
- ☐ 1- or 2-L plastic bottle with cap
- ☐ Scissors
- ☐ Rubber bands (2)
- ☐ Two large balloons

Activity 69: Picture This
- ☐ Vocabulary flash cards
- ☐ Paper or wipe board marker

Activity 71: Incentive Spirometer
- ☐ Bendable drinking straw
- ☐ Scissors
- ☐ Tape
- ☐ 1-inch foam ball or ping pong ball
- ☐ Cardstock, 1 sheet

Activity 74: It's Just a Game
- ☐ Glue
- ☐ Tape
- ☐ Butcher paper or large white paper
- ☐ Markers
- ☐ Ruler
- ☐ Old board game and parts
- ☐ Index cards

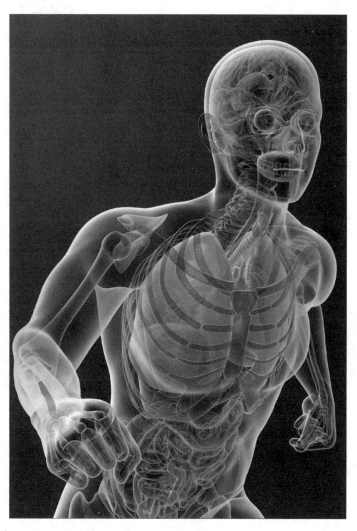

The Breathtaking Respiratory System

Activities & Worksheets
for Use with
*Elementary Anatomy: Nervous,
Respiratory, and Circulatory Systems*

Page left intentionally blank.

| The Breathtaking Respiratory System | After Pages 76–80 | Day 72 | Activity 46 | Name |

The Breathtaking Respiratory System Flash Cards

Carefully cut the vocabulary cards along the dashed lines. Cards are used in multiple activities, so please store in an envelope or secure with a rubber band.

Allergens	Allergy
Alveoli	Anosmia
Apneustic Center	Antiseptic
Asthma	Bacteria
Bronchi	Bronchioles

High acquired sensitivity to certain substances, such as drugs, pollens, or microorganisms, that may include such symptoms as sneezing, itching, and skin rashes	A foreign substance, such as mites in house dust or animal dander, which, when inhaled, causes the airways to narrow and produces symptoms of asthma
Loss of the sense of smell	Small air sacs in the lungs that give the tissue a honeycomb appearance and expand its surface area for the exchange of oxygen and carbon dioxide
Capable of preventing infection by inhibiting the growth of bacteria	The neurons in the brain stem controlling normal respiration
Organisms not able to be seen except under a microscope, found in rotting matter, in air, in soil, and in living bodies, some being the germs of disease	Disease of the lungs characterized by episodic airway obstruction caused by extensive narrowing of the bronchi and bronchioles
Any of the small, thin-walled tubes that branch from a bronchus and end in the alveolar sacs of the lung	The two branches of the trachea that extend into the lungs

| Carbon Dioxide | Chemoreceptors |

| Cilia | Cribriform Plate |

| Cystic Fibrosis | Diffusion |

| Epidemic | Epiglottis |

| Epithelium | Exhale |

| Influenza | Inhale |

A sensory nerve stimulated by chemical means	A colorless, odorless, incombustible gas, formed during respiration
Located in the ethmoid bone of the skull in the nasal cavity where the nerve endings of the sense of smell are found	Short, hairlike, rhythmically beating organelles on the surface of certain cells that provide mobility
The movement of atoms or molecules from an area of higher concentration to an area of lower concentration	An inherited disorder of the exocrine glands affecting mainly the pancreas, respiratory system, and sweat glands
The thin elastic cartilaginous structure located at the root of the tongue that folds over the glottis to prevent food and liquid from entering the trachea during the act of swallowing	An outbreak of a disease or illness that spreads rapidly among individuals in an area or population at the same time
To breathe out	Any tissue layer covering body surfaces or lining the internal surfaces of body cavities, tubes, and hollow organs
To breathe in	A highly contagious viral disease inflammation of the respiratory passages

Iron Lung	Gestation
Goblet Cells	Laryngitis
Larynx	Lower Respiratory Tract
Mucus	Nares
Nasal Turbinate	Organogenesis
Pandemic	Pharynx

The period during which unborn young are "carried" inside the womb	An airtight metal cylinder providing artificial respiration when the respiratory muscles are paralyzed, as by poliomyelitis
Inflammation of the larynx	Cells in the respiratory tract that produce mucus
Consisting of all the structures in the respiratory tract lying below the larynx	The upper part of the trachea containing the vocal cords; also called *voice box*
An external opening in the nasal cavity of a vertebrate; a nostril	The slimy, viscous substance secreted as a protective lubricant by mucous membranes
The development of bodily organs	Any of the scrolled spongy bones of the nasal passages in man and other vertebrates
The passage that leads from the cavities of the nose and mouth to the larynx (voice box) and esophagus	An epidemic that spreads over a very wide area, such as an entire country or continent

Physiologist	Pleura Sac
Pneumotaxic Center	Polio
Stethoscope	Sinuses
Surfactant	Trachea
Tracheostomy	Upper Respiratory Tract
Ventilation	Virus

A membrane that encloses each lung and lines the chest cavity	Biologist specializing in physiology (the biological study of the functions of living organisms and their parts)
Poliomyelitis, an acute viral disease marked by inflammation of nerve cells of the brain stem and spinal cord that can affect the ability to walk and breathe	A nerve center in the upper pons of the brain stem that rhythmically inhibits inspiration
A cavity or hollow space in a bone of the skull, especially one that connects with the nose	An instrument for listening to the sounds made within the body
A thin-walled, cartilaginous tube descending from the larynx to the bronchi and carrying air to the lungs; also called *windpipe*	Surfactant reduces the surface tension of fluid in the lungs and helps make the small air sacs in the lungs (alveoli) more stable
Composed of the parts of the upper respiratory system: the nose, sinuses, pharynx, and larynx	Surgical construction of an opening in the trachea in the front of the neck
Any of various extremely small, often disease-causing agents	The exchange of air between the lungs and the environment, including inhalation and exhalation

| The Breathtaking Respiratory System | After Pages 76–80 | Day 73 | Activity 47 | Name |

Snot O' Matic

CAUTION: Using food coloring or dyes can stain hands and clothing. Protect surfaces as well. Be very careful and ask an adult for help when using it.

Materials:

- Borax Laundry Booster
- 1 liter or quart soda bottle, clean
- Measuring cup
- Green or neon food coloring
- Tap water
- Tablespoon
- White glue

Instructions:

1. Mix approximately ⅛ cup of Borax with approximately ½ liter of warm water in the bottle. Continue to shake the bottle until the borax is dissolved in the water.
2. Let the solution cool to room temperature.
3. Place 4 tablespoons of white glue in the cup.
4. Add drops of food coloring until the desired color is achieved. Stir well.
5. Measure 3 tablespoons of Borax solution from the pop bottle and add it to the glue mixture in the cup. Stir.
6. Now enjoy a little gross fun!

| The Breathtaking Respiratory System | After Pages 81–83 | Day 75 | Worksheet 27 | Name |

The Word of God

Look up Psalm 150:6, and write it below.

_____
_____
_____
_____
_____
_____
_____
_____

What does this passage mean to you?

_____
_____
_____
_____
_____
_____
_____
_____
_____

The Breathtaking Respiratory System	After Pages 84–87	Day 76	Activity 48	Name

Wind Bag

This experiment will measure how much air a person can exhale in one breath.

Materials:

- Several large balloons (one for every one of your test subjects)
- One 2-foot length of yarn or other string
- A ruler
- Markers

Procedure:

1. Ask your test subject (a classmate or family member) to blow into a balloon with one continuous long breath. When the subject cannot blow any more, quickly pinch the neck of the balloon so that no air will escape.
2. Have your subject wrap the string around the widest area of the balloon. Measure the string with the ruler. The length of the string will be the circumference of the balloon (the distance around).
3. Using the graph on the next page, record your results.
4. Repeat the same procedure with several other subjects.

Wind Bag Graph:

INCHES	MARY					
24						
22						
20						
18						
16						
14	■					
12	■					
10	■					
8	■					
6	■					
4	■					
2	■					

(Example: Mary's balloon circumference was 15 inches.)

Test Subjects

Question to Ponder:

Is there a difference in the circumference of the balloon based on the person's size, age, or whether or not the person is a boy or girl?

| *The Breathtaking Respiratory System* | After Pages 84–87 | Day 77 | Worksheet 28 | Name |

Timeline Shuffle

Cut out the following images and paste them in the appropriate order on the timeline.

Joseph Priestley

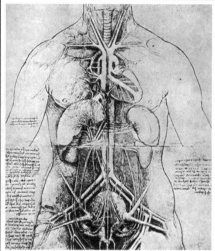
Leonardo da Vinci drawing

Marcello Malpighi

Humphry Davy

Dr. Dorothy Andersen

Evangelista Torricelli

Page left intentionally blank.

Page left intentionally blank.

Tape the timeline pages together

- 470 BC
- AD 170
- AD 1500
- AD 1643
- AD 1660
- AD 1772–1774

Worksheet 28, Day 77 // 151

Page left intentionally blank.

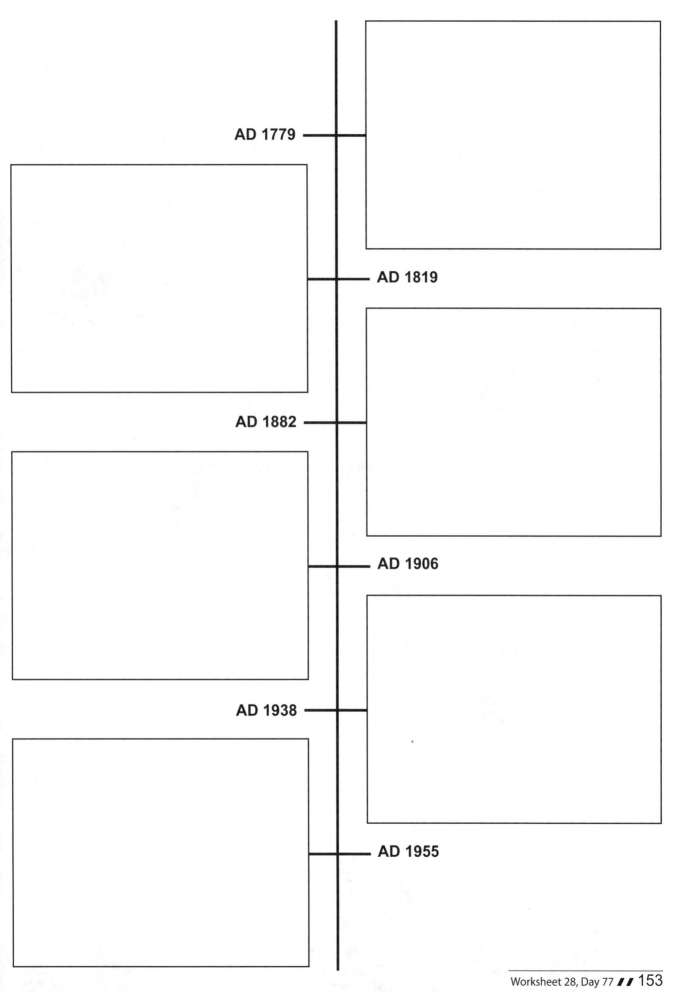

Page left intentionally blank.

 *The Breathtaking Respiratory System* | After Pages 88–89 | Day 78 | Worksheet 29 | Name

Biblical References

Copy the following verse:

John 20:22 — And with that he breathed on them and said, "Receive the Holy Spirit."

Acts 17:25 — And he is not served by human hands, as if he needed anything. Rather, he himself gives everyone life and breath and everything else.

The Dissenter and the Sufferers

Fill in the Blank

1. _____ published six volumes of *Experiments and Observations on Different Kinds of Air* between 1772 and 1790.

2. The diverse group of people that included Baptists, Lutherans, Methodists, Presbyterians, and Quakers that disagreed with the Church of England and broke away were known as _____.

3. As a young boy, Priestley had a ravenous appetite for the _____ and _____.

4. Priestley is credited with discovering several gases including _____ and _____.

5. England's official church was the _____ or Church of England.

6. Today, scientific research is managed and monitored by the government agency called the _____.

7. Many doctors or scientists used to do unsafe things to themselves in order to _____ and learn new _____ advances.

8. _____ are commonly used in laboratory studies.

9. John Haldane was a Scottish _____ born in the late 1800s.

10. The Haldanes' family motto was one word: _____.

Bonus question: How did the experiments of Joseph Priestley and the Haldanes differ?

The Adventures of Jack Haldane

It would seem that Jack Haldane had an interesting life, helping his father with experiments related to air quality and gases. Step back in time and write a first hand account from the perspective of Jack Haldane about one of the experiments you learned about in *The Breathtaking Respiratory System*. You can think of it as a diary entry or as a personal letter he is sending to one of his friends. Use your imagination — include how you think he felt, what the experience might have been like, and what it was like to work with his father on these important studies.

_____
_____
_____
_____
_____
_____
_____
_____
_____
_____
_____
_____
_____
_____
_____
_____
_____
_____
_____
_____
_____
_____
_____

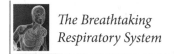

The Breathtaking Respiratory System | After Pages 90–91 | Day 81 | Worksheet 31 | Name

The Basics of the Respiratory System

Fill in the associated boxes telling the parts and functions in the processes of the two parts of the respiratory system.

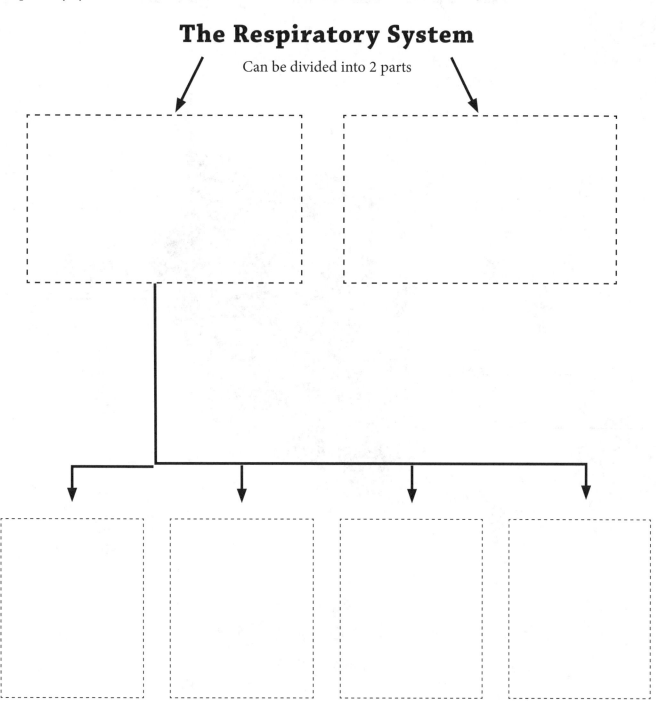

Identification of Parts of the Nasal Cavity

Label the diagram to identify the parts of the nasal cavity and upper respiratory system.

The Upper Respiratory System

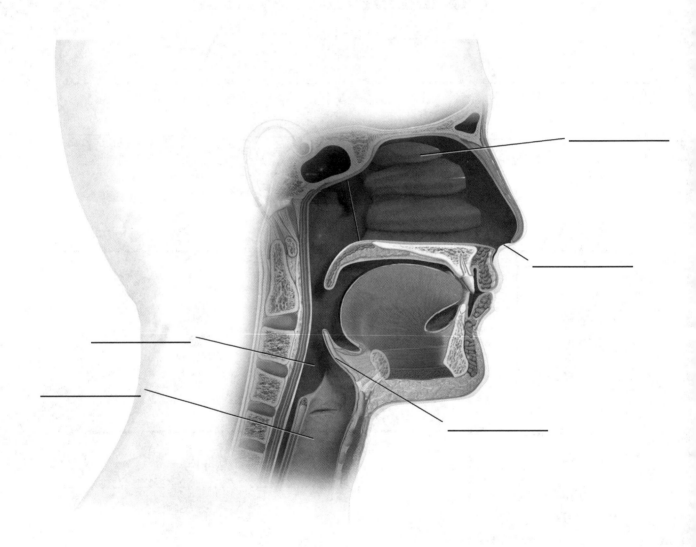

Measuring Lung Capacity

Materials:

Large bowl of water
Milk jug or 2 L pop bottle, marked
Straws

Instructions:

1. Fill the bowl with water.

2. Turn the milk jug/bottle upside-down and submerge it in the water. Hold the cup down and allow the water to completely fill the measuring cup.

3. Place the straw into the water. Make sure the tip of the straw is under the opening to the milk jug/bottle.

4. Blow through the straw into the measuring cup. Air bubbles will begin inside the measuring cup.

5. Continue blowing as long as you are able. Record the measurement on the cup.

6. Once you recover from blowing into the measuring cup, run in place for 1 minute.

7. Repeat the test. How does the exercise affect your lung capacity?

Identification of Parts of the Respiratory Tract

Write the name of each part of the respiratory tract in the blanks below the diagram.

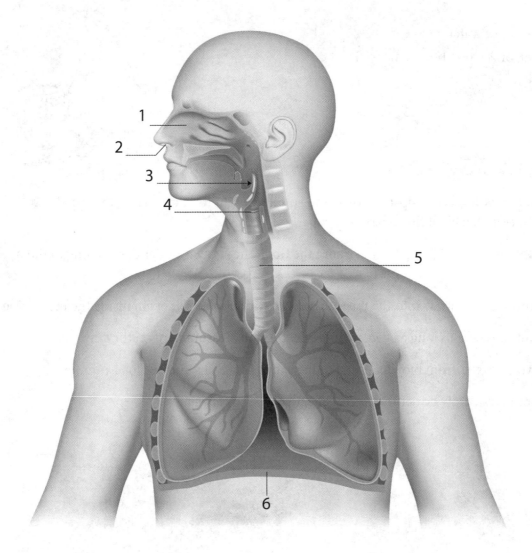

1 _____
2 _____
3 _____
4 _____
5 _____
6 _____

The Breathtaking Respiratory System | After Pages 94–95 | Day 85 | Worksheet 34 | Name

How We Breathe

Review the diagram on the previous sheet. Now, write the process of breathing using the words for the labeled diagram on the lines below.

_____
_____
_____
_____
_____
_____
_____
_____
_____
_____
_____
_____
_____
_____
_____
_____
_____
_____
_____
_____
_____
_____
_____
_____

Build A Word

| ynx | tory | vo | res | phar | tory | co |
| am's | co | ple | vo | dif | upp | wer |
| ba |

Combine the parts of the words to create the words you learned in this part of the book:

1. _____ynx
2. Lar_____
3. _____pira_____
4. _____cal
5. _____rds
6. Ad_____
7. Ap_____
8. Lo_____
9. _____er
10. _____fen_____chia

| The Breathtaking Respiratory System | After Pages 96–97 | Day 87 | Worksheet 36 | Name |

Don't Be Cross!

Solve the clues below then search for the clues in the puzzle.

```
A F E B A C T E R I A B D S D P Q S A T W
N G P E J L B C I L I A U U T V X Y Z K B
O D I F F U S I O N C R M N A B T C S H A
S W D I C O I H A O I L O P L H L R E N T
M I E D N E N V E V R I L A M J E S U S B
I X M R N A U Y K J T O P A N D E M I C C
A S I M E N S P L A R Y N X C H R I S T P
C R C F G M E N T Q R T N A T C A F R U S
A O T G O D S S A V I O R C R O S S T O I
R T R A L L E R G E N S I A M W E O N E S
B P A T T G P J E H O V A H Z R U U V X E
O E C K I P I N F L U E N Z A T R U E M N
N C H H W S T Y B C I H C N O R B Q L U E
D E E E L A H N I L L S P I R I T U A C G
I R A T T C E F G H O H O L Y G O D R U O
O O I R O N L U N G E P I G L O T T I S N
X M F Y F A I T H S V L A R Y N G I T I A
I E T R U E U O Y S L Y A H W E H Q X M G
D H H O P E M E X H A L E P R A Y E R T R
E C V O C A L C O R D S A B B A L W P T O
```

1. _____ A colorless, odorless, incombustible gas, CO_2, formed during respiration
2. _____ A thin-walled, cartilaginous tube descending from the larynx to the bronchi and carrying air to the lungs; also called windpipe
3. _____ Reduces the surface tension of fluid in the lungs and helps make the small air sacs in the lungs (alveoli) more stable
4. _____ A cavity or hollow space in a bone of the skull, especially one that connects with the nose
5. _____ A foreign substance, such as mites in house dust or animal dander, which, when inhaled, causes the airways to narrow and produces symptoms of asthma
6. _____ Loss of the sense of smell
7. _____ Any of the small, thin-walled tubes that branch from a bronchus and end in the alveolar sacs of the lung

8. _____ A sensory nerve stimulated by chemical means
9. _____ Short, hairlike, rhythmically beating organelles on the surface of certain cells that provide movement of fluids
10. _____ The movement of atoms or molecules from an area of higher concentration to an area of lower concentration
11. _____ Any of various extremely small, often disease-causing agents consisting of a particle (the virion), containing a segment of RNA or DNA within a protein coat known as a capsid
12. _____ The two folded pairs of membranes in the larynx (voice box) that vibrate when air that is exhaled passes through them, producing sound
13. _____ An airtight metal cylinder enclosing the entire body up to the neck and providing artificial respiration when the respiratory muscles are paralyzed, as by poliomyelitis
14. _____ Organisms not able to be seen except under a microscope, found in rotting matter, in air, in soil, and in living bodies, some being the germs of disease
15. _____ An acute viral disease marked by inflammation of nerve cells of the brain stem and spinal cord that can affect the ability to walk and breathe
16. _____ An epidemic that spreads over a very wide area, such as an entire country or continent
17. _____ An outbreak of a disease or illness that spreads rapidly among individuals in an area or population at the same time
18. _____ The thin elastic cartilaginous structure located at the root of the tongue that folds over the glottis to prevent food and liquid from entering the trachea during the act of swallowing
19. _____ To breathe out
20. _____ A highly contagious and often epidemic viral disease characterized by fever, tiredness, muscular aches and pains, and inflammation of the respiratory passages
21. _____ To breathe in; inspire
22. _____ The period during which unborn young are "carried" inside the womb
23. _____ The upper part of the trachea in most vertebrate animals, containing the vocal cords
24. _____ The slimy, viscous substance secreted as a protective lubricant
25. _____ The development of bodily organs

Bonus Words				
God	Jesus	Savior	Cross	I AM
True	Hope	Abba	Jehovah	Spirit
Holy God	Faith	Christ	Yahweh	Prayer

| The Breathtaking Respiratory System | After Pages 98–99 | Day 88 | Activity 51 | Name |

Smoke in Your Lungs

Materials:

Spray bottle with water
Cake or egg dye
1 bag of cotton balls
A large straw

Instructions:

(**CAUTION**: Use of food coloring, cake or egg dye can stain hands and clothing. Be careful when using.)

1. You will need a spray bottle with water, and one color of either cake or egg dye.
2. You can poke holes in the straw, and then place it down into the clear bottle with the cotton balls.
3. Place the dye into the spray bottle, and then position the nozzle of the spray bottle at the end of the straw.
4. Spray the bottle a number of times into the straw. Watch as the water with the dye seeps into the cotton balls. Imagine that it is the tar and unhealthy chemicals that are found in cigarettes.

Note: Even something as simple as a spray bottle of water sprayed on a window screen can demonstrate visually the concept of things being trapped in the lungs.

Page left intentionally blank.

The Life of an Oxygen Molecule in the Respiratory System

Choose the word from the word bank that best completes each of the blanks below.

Word Bank				
alveoli	anosmia	bronchioles	carbon dioxide	capillaries
cribriform plate	larynx	nostrils	nasal turbinate (concha)	nasopharynx
oropharynx	right bronchus	tongue	trachea	vocal cords

Once upon a time there lived Siamese twins named Octavius and Ollie. They were happy oxygen molecules. They were traveling on their happy-go-lucky way, minding their own business, when suddenly they collided with an appreciative nose. They entered the nose through the _____. Entering the atrium of the nose, they noticed three slippery slopes called the _____ _____. They churned and turned by those three bony plates. At the top of the passage, they noted a special tissue along the _____ _____ designed to allow the sense of olfaction (smell). Damage to the bones in this area can cause a condition called _____. Tumbling along, they entered a large cavernous space called the _____. Octavius and Ollie descended down this large passageway. Gazing to their left they noted a large space with a muscular appendage called the _____. At this point Ollie was sure they had entered the _____. They came to a fork in the road.

One road descended downward as a muscular tube. At the entrance of the other passageway there were two fibrous stiff flaps called the _____ _____. The doorway to this passage welcomed their entrance. With great curiosity they entered the _____ and traveled down a great wind chamber. They began to be jostled about as their small molecular bodies came across several speed bumps in the _____. The passageway divided again. They arrived at another fork in the road. Tossing a coin, they decided to veer right into the _____ _____. The course continued to branch and narrow. Octavius was really beginning to enjoy his journey, and his anticipation grew. With great delight, they entered the smallest passageways called the _____. Ollie knew their destiny was going to at last be fulfilled! They entered a diminutive balloon-like structure called the _____. The brothers were excited to be greeted by other friends: Corbin, Osmond, and Otto, who also happened to be Siamese brothers. They collectively liked to be referred to by the name of _____ _____ molecule. They gave Octavius and Ollie a set of high fives as they were leaving. They told Octavius and Ollie about all of their wonderful intrigues while in the body and were sad that the party had to end. Corbin, Osmond, and Otto waved good-bye and began their ascent. Octavius and Ollie squeezed across a thin wall and entered the _____. They were on their way. God's design for their life was going to be fulfilled. They relaxed and traveled with assurance down the great life-giving pathway.

| The Breathtaking Respiratory System | After Pages 100–101 | Day 90 | Worksheet 38 | Name |

What's Your Function?

Identify the functions of the different areas of the respiratory system.

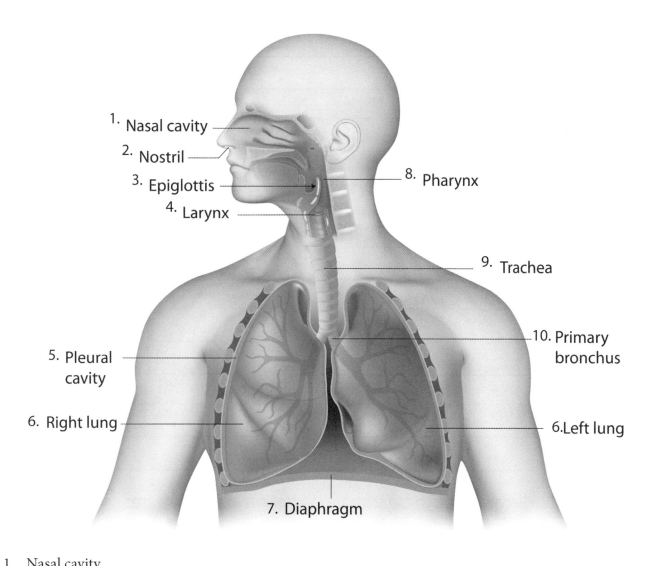

1. Nasal cavity _____

_____

_____

2. Nostril _____

_____

_____

3. Epiglottis _____
_____

4. Larynx _____
_____

5. Pleural sac _____
_____

6. Lungs _____
_____

7. Diaphragm _____
_____

8. Pharynx _____
_____

9. Trachea _____
_____

10. Primary bronchus _____
_____

| *The Breathtaking Respiratory System* | After Pages 102–103 | Day 91 | Activity 52 | Name |

What's the Count?

You are unaware of most breaths you take in a given minute or day. Most of the time, breathing can be an effortless activity. Have you ever wondered how many breaths you may actually take during rest? Let's give this a shot.

Take a timer and place it on the table in front of you. Take your open hand and place it on your chest.

1. Count how many breaths you take in one minute.

2. How many breaths will you take in an hour?

3. How many breaths will you take in an entire day?

4. How many breaths will you take in a week?

5. How many breaths will you take in an average month?

6. How many breaths will you take in a year?

7. The average life expectancy in the United States for a male is about 79 years and the average for a female is 81 years. Based on gender, what would be the total number of breaths you will take in your life?

Inhale, Exhale

Breathing is a basic function of life.

To inhale means to breathe in.

To exhale means to breathe out.

Look at the picture to the right. Use the picture to answer the questions below. Write the words **Inhale** or **Exhale** on the spaces that the line describes.

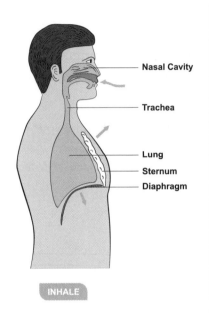

INHALE

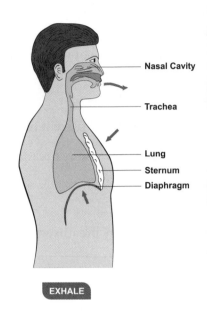

EXHALE

1. The diaphragm moves downward. _____

2. You breathe in. _____

3. Air exits out of your mouth and nose. _____

4. Your body rids itself of carbon dioxide. _____

5. The lungs expand. _____

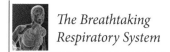

The Breathtaking Respiratory System | After Pages 106–107 | Day 93 | Activity 53 | Name

The Voice of a Balloon

Materials: Balloon

Instructions:

1. Blow up a balloon.
2. Pinch the neck of the balloon closed.
3. Stretch the neck of the balloon tight and allow the air to slowly escape. Vary the tension across the neck of the balloon.

Compare the differences in pitch with male and female voices.

Breathe

Have you ever noticed that when you breathe your chest moves?

When you inhale, your chest expands outward and air rushes into your lungs. In this experiment, you will observe what happens when you breathe.

Materials:

Paper lunch bag

Procedure:

1. Find a nice soft place to lie down on your back.
2. Take the paper bag with one hand and hold it over your mouth. Place your other hand on your chest.
3. Breathe slowly and deeply.
4. Observe what happens to the bag when you breathe in and when you breathe out.

Questions:

1. What happens to the bag when you breathe?

_____

_____

_____

2. What does your other hand feel on your chest as you breathe?

_____

_____

_____

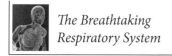

The Breathtaking Respiratory System | After Pages 108–109 | Day 95 | Worksheet 40 | Name

Digging a Bit Deeper

Pick one of the questions below and do a bit of research to explore the answers. Use at least two resources. On the back of this worksheet, explain your findings.

1. The lungs are considered to be a respiratory organ and an excretory organ. Explain why this is so.

   _____
   _____
   _____

2. How is respiration similar to burning fuel?

   _____
   _____
   _____

3. Emphysema is a lung disease that can be caused by smoking. Explain what happens to the alveoli in this disease and what it does to breathing.

   _____
   _____
   _____
   _____

4. What is altitude sickness?

   _____
   _____
   _____

Page left intentionally blank.

| *The Breathtaking Respiratory System* | After Pages 108–109 | Day 96 | Activity 55 | Name |

Wind Bag

Measure lung capacity with this durable, silk-screened plastic bag, calibrated to six liters, with accompanying mouthpiece and holder. Shows the relationship of body size to lung volume. Kits contain four lung bags, four mouthpieces, four mouthpiece holders, and four rubber bands. Grades 3–9. Instructions included.

Question:

Is there any relationship between a person's height and their lung capacity? What do you think?

Steps:

1. Measure the height of each family member and record the results.

Name	Height

2. Attach the cardboard tube to the opening of the bag.
3. Have each person take turns measuring his or her lung volume. Each should stand up, take a long, deep breath, and exhale, then take another breath. Seal lips around the tube, and exhale completely into the bag.
4. Once exhalation is complete, quickly grab the bag below the mouthpiece to trap in the air.
5. Push the trapped air to the distant end of the air bag. The easiest method is by placing the bag over your thigh and pressing the bag against your thigh and pushing the air down.
6. Record the level of the air for each person in the chart.
7. Plot the results on the following graph.

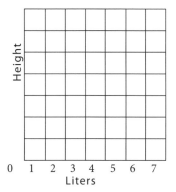

Explain your findings:

What Is Your Girth?

Musical vocalists must train and learn to breathe when they sing. They are told to sing from the diaphragm instead of their vocal cords. What does that mean?

In this experiment, you will need a tape measure. You may need someone to assist you.

1. Place a tape measure around your chest, just below your armpits. Take a deep breath in and measure the results.
2. Exhale completely. Take the measurement in the same place.

INHALATION GIRTH: _____

EXHALATION GIRTH: _____

Is there a difference in the two measurements? To what would you attribute the difference? What is responsible for facilitating this difference?

The Uniqueness of the Lungs Report

1. What makes the lungs unique among the organs being formed while in the womb?

2. What role does the amniotic fluid play related to the lungs?

3. Which phase of lung development do the grape-like clusters of the alveoli develop at the ends of the terminal bronchioles?

4. What phase of lung development do the lung tissue, alveolar sacs, as well as type 1 and type 2 cells form?

5. Having read the detailed process of how the lungs are formed, does it seem to you that it is something that just happens by accident, or that you and your lungs are "wonderfully made"?

| The Breathtaking Respiratory System | After Pages 110–111 | Day 98 | Activity 58 | Name |

A Nose for Clues

Foods come in all kinds of tastes, from the sweet, tangy taste of a tangerine to the salty taste of potato chips. No doubt, our sense of taste contributes to the enjoyment we experience when we eat.

Materials:

A bag with an assortment of flavored hard candies (lollipops will work as well)

Bandana or scarf

Partner

Procedure:

1. Wrap the bandana/scarf around your partner's head, covering the eyes. No peeking.
2. Have your partner pinch his or her nose and breathe out of the mouth.
3. Chose a candy for your partner to taste. Do not mention the flavor or color. Have your partner taste the candy with the nose still pinched.
4. Ask your partner to describe the taste of the candy when it is placed in the mouth. Then allow your partner to keep the candy in his or her mouth for 30 to 60 seconds (still with nose pinched). Observe his or her reaction.

    Did your partner notice any change in the taste? Did time make any difference in the taste?

    _____
    _____
    _____

5. Have your partner now taste the candy without pinching the nose.

    What is the difference?

    _____
    _____
    _____

6. Change places and now you be the taster.
   What significance does smell have on your ability to taste?

Scrambled Eggs, Level 1

Unscramble the mystery words below.

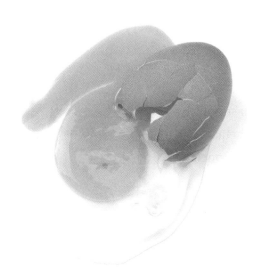

1. snaers _____
2. iialc _____
3. cuusm _____
4. uvris _____
5. aahsmt _____
6. xylrna _____
7. oellvai _____
8. yrleagl _____
9. xlahee _____
10. xyprahn _____

Word Box				
alveoli	asthma	cilia	exhale	larynx
mucus	nares	pharynx	virus	allergy

Scrambled Eggs, Level 2

Unscramble the mystery words below.

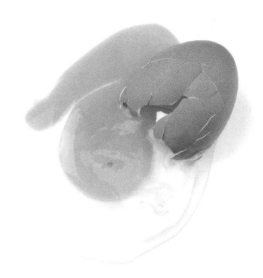

1. ffsdniuo _____
2. aasmions _____
3. chetaar _____
4. dampncie _____
5. ssinsue _____
6. zuinflena _____
7. lelasengr _____
8. eemcipid _____
9. oliop _____
10. latetvionn _____

Word Box				
allergens	anosmia	diffusion	epidemic	influenza
pandemic	polio	sinuses	trachea	ventilation

| The Breathtaking Respiratory System | After Pages 116–117 | Day 102 | Activity 59 | Name |

Grape-Like Clusters

Materials:

Balloons (small)
String
White glue
Yarn (blue, red, yellow)
Rope

Instructions:

1. Blow up the small balloons to the size of a grape. Each of the balloons will represent an alveolus filled with air.
2. Be sure to tie up each of the balloons and attach a piece of yellow yarn to each one to assure no air will escape.
3. Gather up all the balloons and tie each of the free ends of the yarn together.
4. Soak additional strings of blue and red yarn in glue. Then wrap the glue-soaked strings around the balloons.

Scrambled Eggs, Level 3

Unscramble the mystery words below.

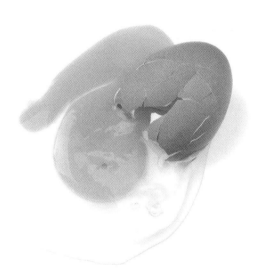

1. stiapenuc teencr _____
2. thempielui _____
3. stagenoit _____
4. gangenesoois _____
5. teempoorchcers _____
6. stoomracheyt _____
7. salan bineratut _____
8. factantsur _____

Word Box			
chemorecepters	epithelium	organogenesis	gestation
surfactant	nasal turbinate	apneustic center	tracheostomy

The Breathtaking Respiratory System | After Pages 116–117 | Day 103 | Activity 60 | Name

Up in Smoke

You can do one of the following activities:

1. Create a small poster encouraging people not to smoke. You can include details from the information you have read in the book and cut out images or text from magazines.

2. Create an anti-smoking song that includes some of the information of why its not good or healthy to smoke. Use at least four of the following phrases in your song:
   Don't smoke
   Be healthy
   Don't start
   Smoking causes lung cancer
   Emphysema
   Asthma
   Your lungs can't work

3. Write a short conversation between two friends — one has a cigarette and is trying to talk the other friend into smoking it too. Have the non-smoking person talk his friend out of smoking by sharing some of the health risks and problems associated with cigarettes.

_____
_____
_____
_____
_____
_____
_____
_____

A		B		C		D		E		
F		G		H		I		J		
K		L		M		N		O		
P		Q		R		S		T		
U		V		W		X		Y		
Z										

Every Breath You Take

The Breathtaking Respiratory System — After Pages 118–119 — Day 105 — Worksheet 44 — Name

Solve the hidden message on other side using the code key above.

When you breathe, you inspire.

___ ___ ___ ___   ___ ___   ___ ___

___ ___ ___   ___ ___ ___ ___ ___ ___ ___

___ ___ ___   ___ ___ ___ ___ ___ ___

| The Breathtaking Respiratory System | After Pages 120-121 | Day 106 | Worksheet 45 | Name |

Lung Damage – The Awful Effects of Smoking

Fill in the right-hand column telling what happens after these stages of stopping smoking.

When this occurs:	Then this happens:
20 minutes after one's last cigarette	1.
8 hours after one's last cigarette	2.
48 hours after one's last cigarette	3.
A few weeks after one's last cigarette	4.
1 year after one's last cigarette	5.
5 years after one's last cigarette	6.
15 years after one's last cigarette	7.

Page left intentionally blank.

| The Breathtaking Respiratory System | After Pages 120–121 | Day 107 | Activity 61 | Name |

Pulse and Respiratory Rates

Materials:

Stopwatch or timer with a second hand

Jump rope

Partner

Instructions:

1. Locate the site of a pulse in your partner. Try locating it in one of two places. You can place two fingers at the angle of the jaw on the neck or the underside of the wrist near the fleshy bottom aspect of the thumb. Be sure not to use your thumb to locate the pulse. (Your thumb has its own pulse, and it may be difficult to determine which pulse you are feeling.)

2. Have a timer on hand. Locate the pulse, and count the number of beats that you feel over the course of a minute. Record the number.

3. Closely watching your partner, count the number of breaths he or she takes during a minute. Record your findings.

4. Now the fun begins. Have your partner jump rope for one to five minutes continuously. If your partner is unable to jump rope, just jump up and down in place for the duration of the time.

5. At the conclusion of the jumping, count the number of beats of your partner's pulse for a minute. Record. Then count the number of breaths he or she takes. Record.

Resting	After Jumping
Pulse rate:	Pulse rate:
Breaths per minute:	Breaths per minute:

Observations:

1. Observe your partner before and after activity. What happens to the pulse and respiratory rate following activity?

	Before Jumping	After Jumping
Pulse		
Respiratory Rate		

2. Explain why this happens. Why would the pulse and respiratory rates be affected by activity?

3. How long does it take for your partner's respiratory and pulse rates to return to their original state?

| The Breathtaking Respiratory System | After Pages 120–121 | Day 108 | Activity 62 | Name |

The Air Around Us

Depending on where you live, whether in the city or country, the air that you breathe can contain various types of pollutants. In this lab, you will examine the air that you breathe.

Materials:

- One sheet of graph paper
- 5 microscope slides*
- Petroleum jelly
- Pencil
- Scissors
- Tape

Instructions:

1. Prepare the instruments that you will use for observation of the air quality. Place the slide on the graph paper and trace the outline of the slide on the paper. Cut the outline of the slide out. Do this for each of the five slides.
2. Tape each piece of graph paper to each of the slides. Be careful to place the tape on the back of the paper and affix it to the slide so that no tape will be hanging on the other side of the slide.
3. Determine five different locations you will place each of the slides. (For example, a shelf, kitchen, basement, or outside in your window, etc.) Label the date on the slides. Record the location of the slides.
4. Prior to placement of the slides, rub a thin layer of petroleum jelly on the opposite side (the side without the affixed graph paper). The jelly will trap and collect pollutants that may be in the air.
5. Leave the slides in place for one to three days.
6. Collect the slides on the same day. The graph paper will allow you to determine your observations. Count the number of particles that you find in 10 of the squares. Record your results below:

Slide	Location	Number of Particles
1		
2		
3		
4		
5		

*If you don't have microscope slides, you can modify the experiment with either a clear cup, glass, or small plate. Or you can use small squares of cardboard and clear food wrap over the squares, taping it together on the back. Continue the experiment. You can also use a magnifying glass to help see the particles if needed. Discuss the findings of this experiment. What does it teach you about the air that we breathe?

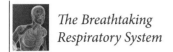 *The Breathtaking Respiratory System* | After Pages 122–123 | Day 110 | Worksheet 46 | Name

Good Habits for Health

When you sneeze or cough you should: (circle all those that are appropriate)

Cough into your hands.

Go to the park.

Touch your face after you sneeze.

Cover your mouth and nose with a tissue.

Leave your used tissue laying around.

Cough into your upper sleeve.

Get more rest.

Rub your hands on your jeans.

Use hand sanitizer.

Wash with water, but no soap.

Touch places that other people will be touching, like the water fountain at school.

Put your used tissue in a trash can.

BONUS: What is the word that means "to sneeze"?

_____
_____
_____
_____
_____
_____
_____

 | *The Breathtaking Respiratory System* | After Pages 122–123 | Day 111 | Activity 63 | Name

No Idling (Hint: Visit a website or check with your local health office to see if they have information available.)

The above pictures are a few signs used in the campaign to discourage idling of engines. Research the concerns that anti-idling campaigns raise.

1. Why may idling engines be a hazard?
   _____
   _____

2. Discuss the various ways that it may harm the environment and health.
   _____
   _____

3. What is your opinion of these types of campaigns?
   _____
   _____

4. Do you think they are helpful?
   _____
   _____

5. Are they effective?
   _____
   _____

6. Why or why not?
   _____
   _____

7. Draw your own anti-idling sign!

What Trees Want

NOTE: Have an adult supervise this experiment.

Gas exchange occurs at the alveolar level in the lungs: one gas, oxygen, is exchanged for another gas, carbon dioxide. Carbon dioxide is a gas that plants need to perform photosynthesis. We cannot see carbon dioxide directly. In this experiment, we have ways to explore this byproduct of our cellular processes in the body.

Materials:

  2 test tubes
  Test tube rack or holder
  Bromothymol blue solution (this can be purchased in the fish supply section of a pet store)
  Straw
  Stopwatch
  Glass-marking pencil (China pencil) or a piece of tape or label
  Jump rope

Procedure:

Part 1

1. Fill each of the test tubes half full of water.
2. Label the test tubes #1 and #2 with glass-marking pencil.
3. Place 3 to 4 drops of bromothymol blue solution in each test tube.
4. Place the straw into test tube #1. Make sure you do not spill the fluid. Blow gently into the straw so the water begins to bubble slightly. **CAUTION: Be sure to ONLY blow — DO NOT suck any liquid into the straw. An adult may need to do this part.**
5. Using the stopwatch, as you begin to blow in the test tube observe the time it takes to see some type of response.

6. Compare the two test tubes. What happened in each of the test tubes? What is the purpose of test tube #2?

Draw a picture of observations below:

Part 2

Caution: Remember an adult may need to do number 5.

1. Empty test tube #1. Refill halfway with water.
2. Next, jump rope or jump in place for 3 to 5 minutes.
3. Fill the test tube half full with water immediately after jumping.
4. Add 3 to 4 drops of bromothymol blue to test tube #1.
5. Blow gently into test tube #1. Record the time it takes to change. Was there a difference in Part 1 and Part 2 of the experiment? Why do you think there was a difference?

Fill in the Blanks

organ   producers   pancreas   fibrosis

disease   defect   mucus   genetic

malfunctions

Cystic _____ is an inherited _____. It is a disease passed from parent to child in their _____ instructions. The complexity and precision with which God designed the body blows our minds. A small addition, change, or deletion in the genetic instructions in the body can cause huge _____. Cystic fibrosis is a result of a specific genetic _____ that interferes with the body's ability to carry salt and water to and from the cells. Due to this defect, many _____ systems in the body are impaired. Thick, gummy _____ builds up and clogs the lungs and the digestive tracts. It affects the lungs, _____, liver, intestines, sinuses, and reproductive organs, the areas that are high _____ of mucus in the body.

| The Breathtaking Respiratory System | After Pages 124–125 | Day 114 | Activity 65 | Name |

Just the Facts: The Spanish Flu Pandemic of 1918

Here is your chance to be a young reporter on the trail of a big story. Use the following facts to create a short 3-paragraph news article using the "who, what, when, where, and why" method.

WHO	WHAT	WHEN	WHERE	WHY
Who was affected?	What was the disease?	When was the pandemic happening?	Where were the places people became sick?	Why did it affect so many in so many places?

Here are the facts; pick and choose the ones to include in your story:

- Those who died from it were normally healthy younger people, instead of the elderly or young children, who were sickest in the first wave of the pandemic.

- No one knows where it started, but the first case in the United States was discovered in Kansas.

- 500 million people of all ages; 50 to 100 million died

- It is thought to have spread worldwide because of troop movements in World War I and also improved methods of travel between countries and continents.

- The flu impacted people quickly, often before the governments could create rules for helping to stop the spread of the virus.

- It was known as the Spanish Flu; it was officially the H1N1 influenza virus.

- Between January 1918 and December 1920

- All around the world, including islands in the Pacific and even the Arctic.

- Entire villages of people in Alaska exposed to the flu died.

- The reason for the end of the pandemic is uncertain.

- People wore masks and closed businesses, and the spread of this flu is believed to have helped impact the end of World War 1.

| The Breathtaking Respiratory System | After Pages 124–125 | Day 115 | Activity 66 | Name |

A Very Simple Lung Model

Materials:

Clear plastic bottle – 1 liter or 2 liter in size
Scissors
Two balloons – medium or large size
Two rubber bands

Instructions:

1. Either hold the bottle between your knees or place it on a table. Cut the base of the bottle completely off (with adult assistance).

2. Take one of the balloons and, while holding on to the top, slip it into the top of the bottle.

3. Stretch the opening of the balloon over the edge of the bottle's opening. Be sure to roll the balloon's edge down the ridges of the bottle's top to secure it. (You can always use a rubber band to help fasten it, but don't use it too tightly or it will cut into the balloon.)

    TIP: Try to center the opening of the balloon in the center of the bottle's opening. If the balloon should end up loose in the bottle, you can turn up the bottle and fish it out with your finger. Or you can use tweezers to pull it out of the opening and try again.

4. Gently squeeze the middle of the bottle. What do you see? Now place your hand gently over the top of the bottle, leaving a little space for air. That might help – but you should see the balloon expand and contract like your lungs do.

5. Cut the tip of the other balloon, then tie the other side in a knot. Secure it tight over the base of the bottle that you cut off. Pull on the knotted end and watch the other balloon inflate inside the bottle.

Now, let's record some observations from your simple lung model on the following page!

Breathing happens so naturally and so many times a day, sometimes we don't even notice! This activity helped to highlight how your lungs function.

What was the most difficult part of putting your lung model together?

_____
_____
_____
_____
_____
_____
_____

What was the most interesting thing you learned?

_____
_____
_____
_____
_____
_____
_____

| The Breathtaking Respiratory System | After Pages 126–127 | Day 117 | Worksheet 48 | Name |

Puzzle Fun

Can you help the doctor reach the patient?

In your opinion, what are some special challenges that doctors and patients have in rural areas away from large cities and medical centers? Try to write at least three challenges.

Can you reach the sinuses and nose?

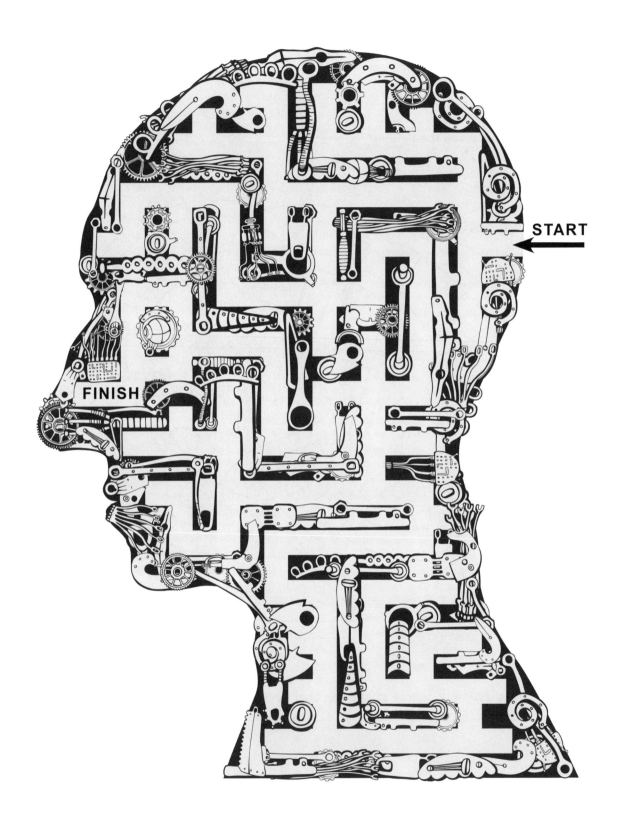

Know Your Noses

Match the animals listed to their noses below:

1. camel
2. cow
3. chimpanzee
4. pig
5. dog
6. seal
7. cat
8. mandrill baboon
9. box turtle
10. elephant
11. koala
12. proboscis monkey

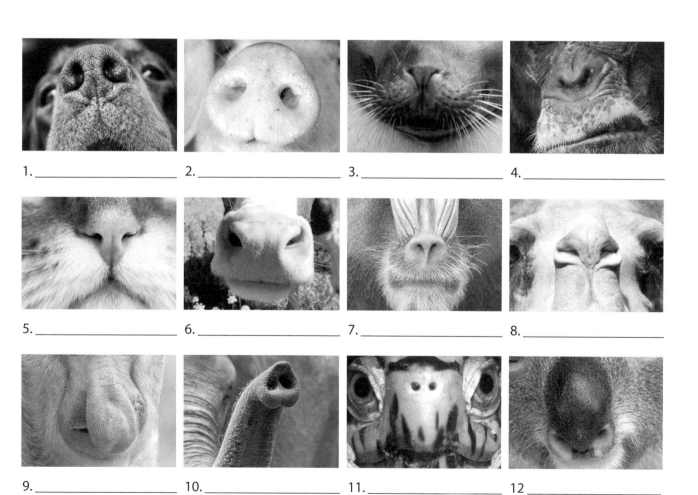

1. _____
2. _____
3. _____
4. _____
5. _____
6. _____
7. _____
8. _____
9. _____
10. _____
11. _____
12. _____

Respiratory Problems

Circle the words that are the name of a respiratory illness or cause of an allergy:

When Asthma Attacks

Put the following three steps in the order they occur in an asthma attack by numbering them 1–3:

_____ The tiny straps of smooth muscle that encircle the bronchioles spasm and tighten around the tubes.

_____ The lining of the airways becomes inflamed and swollen. This swelling narrows the passageway.

_____ The lungs increase production of mucus. This mucus clogs the airways, further narrowing the area.

The Breathtaking Respiratory System | After Pages 128–129 | Day 122 | Activity 68 | Name

Sticky Situations

As we grow, we all encounter situations that may at times be uncomfortable and difficult to navigate. Some of these "sticky situations" may occur in the company of friends. Read the following scenarios. Think about how you would respond to the situations and what you would say. Discuss your answers with a family member or friend.

Scenario #1: You are walking home from basketball practice and you notice a few of your good friends up ahead. As you draw closer to them, you catch the strong smell of cigarette smoke. "Hey guys, what are you up to?" you ask innocently. A few nervous giggles emanate. Your friends quickly make attempts to hide the cigarettes. What do you do and say?

Scenario #2: Your aunt has been smoking for years. She decided at the beginning of the year to make a New Year's resolution to stop smoking. She has attempted to stop smoking in the past but was unsuccessful. What advice would you give to help her be successful this time?

Scenario #3: You are asked to speak to a group of aliens who recently landed on earth. They would like to experience the earth custom of smoking. Explain to them why this is not a healthy habit and why people persist in it despite knowing the hazards.

Picture This

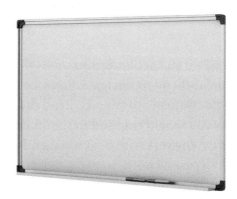

Materials:

Vocabulary flash cards
Paper or wipe board and marker
Partner(s)

Procedure:

1. Draw a card from the deck of vocabulary words. Make sure no one else sees the card.
2. Without talking, draw a picture of the word. See if anyone can guess which vocabulary word is illustrated.
3. The correct guess is awarded a point. The first team or person to collect 10 points wins.

The Doctor Is In

Being a doctor is much like being a detective. You are presented with a situation and you must ask many questions, look at the clues, and perform an examination to discover the answer to the mystery. Doctors are trained to be organized in their presentation of details in regard to their patients. One way a doctor condenses all the information about a patient is in the form of a SOAP note. No, it isn't a note written in soap. SOAP is an acronym for Subjective, Objective, Assessment, and Plan. Subjective means the personal account that the patient or parent uses to describe what they have seen happening.

This can be influenced by emotions and feelings. Objective is the presentation of the measureable facts. It is what is measured and seen directly at the time of the examination. Assessment is the analysis or summation of the problem. Last, the plan is the set of actions to treat the problem.

In this scenario, you are the doctor and you are presented with your first patient at God's Wondrous Machine General Hospital. Your objective is to assess the patient and devise a treatment plan.

Hey, Doc, here are a couple of pointers prior to entering the room to examine your patient. Doctor "talk" is very specific so that all medical personnel understand what is being communicated. Here is your "cheat" sheet:

- Normal values:
   Temperature: 98.6°F
   Heart Rate = 60–100 beats per minute
   Pulse-ox is the amount of oxygen saturation of the red blood cells. Normally, for a healthy person you would like to see the saturation 94% or higher.

- TM = tempanic membrane (eardrum): when the membrane is infected, the membrane may not move easily.

- Nasal flaring occurs when people are attempting to bring additional air into their noses because they are short of breath. The nostrils flare.

- The ratio between the inspire and expire time is important to document. Normally, it takes far less time to breathe out than in. Take a breath now. Do you see that normally it is much quicker to breathe out? In conditions like asthma, air is trapped in the lungs and the expiratory time is prolonged.

- Subcostal retractions: this is the appearance of the ribs as the area under the ribs pulls in when breathing. People experience retractions in this area when they are trying to force more air into their lungs.
- Tachycardia = fast heart rate
- URI = upper respiratory infection, usually viral in nature
- Exacerbation = worsening symptoms

Now go to the Progress Report below and choose two questions from numbers 1–6 to answer on the next page.

GENERAL HOSPITAL

PROGRESS REPORT

Name: Ben Abreeze

Patient ID: 112233

DATE		
6/10/2023		
	S:	10-year-old male, with past medical history significant for recurrent colds and prior hospitalizations for shortness of breath at 6 months and 6 years of age, presents with shortness of breath, cough, and wheezing for the last 2 days. Mother states child began with cold-like symptoms four days prior to the onset of wheezing. Symptoms have consisted of low-grade fever, clear nasal discharge, dry unproductive cough, decrease oral intake of food, and decreased activity.
	O:	**Temp.** = 100.1; **Resp. rate** = 30
		Heart rate = 130; **Blood pressure** = normal limits
		Pulse-ox = 93% on room air
		Ears: slight redness of the TM, mobile
		Nose: + clear nasal discharge + nasal flaring
		Throat: slightly injected (red), no exudates (pus)
		Lungs: + expiratory wheeze heard most at the base of the lungs, Good air entry
		+ prolonged expiratory phase + subcostal retractions
		Heart: + tachycardia, regular rhythm
	A:	10-year-old male with URI and acute exacerbation of asthma.
	P:	What is your plan, Doctor?

Use at least 2 to 3 resources (Internet, books, video, etc.) to answer the following questions. Pick two of the following to explain.

1. How can a cold contribute to the worsening of the symptoms for asthma?

2. Explain to the child and mother what asthma is.

3. What are some of the triggers for asthma?

4. How is asthma treated? What traditional or homeopathic measures are used?

5. What other conditions are associated with asthma?

6. Interview a person who suffers from asthma. What does he or she feel when he or she has an asthma attack? What does he or she use to control his or her symptoms? How long has he or she known that he or she has asthma?

7. Write a short story (at least 12 sentences) of when you may have been sick and had trouble breathing. Did you have to go to the doctor or take medicine?

_____
_____
_____
_____
_____
_____
_____
_____
_____
_____
_____
_____

8. What helped you when you were sick — humidifiers, inhalers, etc.?

_____
_____
_____
_____
_____
_____
_____

| The Breathtaking Respiratory System | After Pages 130–131 | Day 126 | Activity 71 | Name |

Incentive Spirometer

After surgery, it can be painful or difficult to take in deep breaths. Many patients will avoid doing so because of the pain. Unfortunately, this can lead to infections in the lungs to complicate the patient's recovery. Many times an incentive spirometer is given to patients to encourage them to take deep breaths.

Materials:

- Bendable drinking straw
- Scissors
- Tape
- 1-inch foam ball or ping pong ball
- Card stock, 1 sheet

Procedure:

1. Trace a circle approximately 4 to 5 inches across on the piece of cardstock.
2. Cut out the circle. Cut a slit, stopping at the middle.
3. Make a funnel-like shape out of the paper and tape the edges.
4. Snip a small section off the bottom of the funnel, only big enough to slide the straw through.
5. Tape the funnel to the straw. Be sure to leave at least ¼ to a ½ inch of the straw sticking up inside of the funnel for the experiment to work correctly.
6. Once the straw is taped in place, you can use a small amount of putty or clay to seal inside the funnel with the straw (optional). Be careful not to block the center of the straw.
7. Place your ball into the funnel and blow through the other end. See how high and how long you can keep the ball up in the air.

TIP: This can be done without the funnel, but it could take a few tries and a strong, sustained breath!

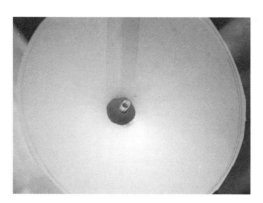

Results of experiment:

It's a Secret!

Can you solve the code to find who this person is?

Short Answer:

1. Who was this person?
   _____
   _____

2. What was the illness that left his legs paralyzed?
   _____
   _____

3. Today, the name of the charity he helped to start is called what?
   _____
   _____

4. Why did this person not tell the American people about his disability?
   _____
   _____

5. When was the first iron lung created?
   _____
   _____

6. Did everyone who used an iron lung have to use it for the rest of their life?
   _____
   _____

7. How expensive could an iron lung be in the 1930s?
   _____
   _____

8. If one dollar in 1930 is equal to $14.25 today, what would the cost of the 1930s iron lung be today?
   _____
   _____

The Breathtaking Respiratory System | After Pages 134–135 | Day 129 | Worksheet 51 | Name

Martha and Wilma

You have now read about the lives of Martha Mason and Wilma Rudolph. At first it seems there is little in common between the two women beyond the fact that polio impacted their lives. "It is not what we lack but what we do with what is gifted to us." Write a short essay on these women with this thought in mind and the achievements of their lives. Or imagine the two women had a chance to meet — what do you think they would talk about?

Sing Your Heart Out

Another way to help you retain new words and lock them in your head is to sing a familiar song. Take a song you learned when you were little, like "Twinkle, Twinkle, Little Star" or "Three Blind Mice," and change or add words from your vocabulary list and sing your heart out.

 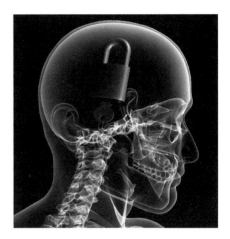

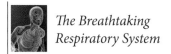

The Breathtaking Respiratory System | After Pages 136–137 | Day 131 | Activity 73 | Name

It's a Record!

Either go to the library to research or search online (with adult supervision) for at least three world records related to the respiratory system functions which can include sports, medical conditions, and even unusual records for longest time sleeping, etc. Give a brief description of each of the three including the year the record was set, who is the record holder, and where it took place.

1. _____

2. _____

3. _____

The Breathtaking Respiratory System | After Pages 138–139 | Day 132 | Activity 74 | Name

It's Just a Game

Games are always a great deal of fun. Have you ever designed your own? Games are an excellent way to learn new concepts. In this activity you will design your own game.

Materials:

- Glue
- Tape
- Butcher paper or large white paper
- Markers
- Ruler
- Old board game and parts (you can purchase inexpensive board games and parts from a second-hand store)
- Index cards

Procedure:

1. Take the old playing board and cover it completely with butcher paper or any large sheet of paper.
2. Glue the paper in place.
3. Design a route on your board with a start and finish.
4. On the index cards, write out an assortment of questions on the respiratory and nervous system. Players are allowed to progress on the board as they answer the questions correctly.
5. Be creative — make up your own rules and have fun!

| The Breathtaking Respiratory System | After Pages 140–141 | Day 133 | Worksheet 52 | Name |

Ponder This!

You read the answers, but can you explain it in your own words? Discuss the following questions with your teacher.

1. Why do I hiccup?

2. Why does my nose run when I cry?

3. Why is snot green?

4. What causes nosebleeds?

5. Why is it hard to breathe at high altitudes?

Word Match

Match the word with its definition:

A. Neutrophils

B. H_2O

C. Epistaxis

D. Myeloperoxidase

E. Spiracles

_____ Another word for a nosebleed.

_____ White blood cells at the site of a bacterial attack.

_____ Represents the composition of water.

_____ Special enzymes that help to destroy bacteria.

_____ Holes that allow air to enter the insect's body.

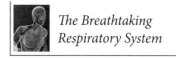

| The Breathtaking Respiratory System | After Pages 142–143 | Day 135 | Worksheet 53 | Name |

Do You Know?

We have taken quite a journey through *The Breathtaking Respiratory System*. Write down five facts that you learned during your study of the respiratory system.

1. _____
   _____
   _____
   _____

2. _____
   _____
   _____
   _____

3. _____
   _____
   _____
   _____

4. _____
   _____
   _____
   _____

5. _____
   _____
   _____
   _____

Page left intentionally blank.

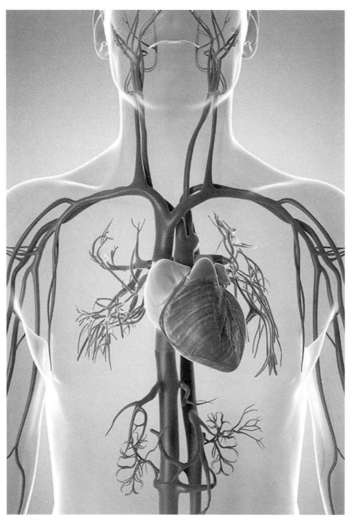

The Complex Circulatory System

Educator Aids
for Use with
*Elementary Anatomy: Nervous,
Respiratory, and Circulatory Systems*

Page left intentionally blank.

The Complex Circulatory System

CIRCULATORY SYSTEM OBJECTIVES

Successful completion of this module will enable the student to:

- Identify at least three themes from the Bible demonstrating the significance of our "hearts."
- Identify the functions of blood.
- Name the different parts of the blood.
- Explain the function of each blood cell.
- Explain the basic process of a blood clot formation.
- Identify the basic blood types.
- Explain the donor/recipient relationship between the various blood types.
- Compare and contrast the similarities and differences between arteries and veins.
- Name the five types of vessels in the body as blood flows through the complete circuit from the left side of the heart around the body to the right side of the heart.
- Explain the difference between open and closed circulatory systems.
- Locate and label structures of the heart in a diagram.
- Trace the pathway of blood in the heart using arrows on the diagram.
- Recognize the attributes of the heart structure as a double pump system (right and left side).
- Define the terms systole and diastole.
- Discuss heart sounds in terms of what they represent and how they sound.
- Discuss how to keep the cardiovascular system healthy.

Page left intentionally blank.

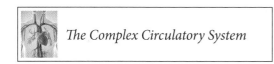
The Complex Circulatory System

SUPPLY LIST FOR ACTIVITIES

Activity 75: Circulatory Flash Cards
- [ ] Scissors
- [ ] Tape or glue stick

Activity 78: Blood Splatter
- [ ] Paints
- [ ] Paint brush
- [ ] Toothbrush
- [ ] Glitter
- [ ] Newspaper
- [ ] School glue
- [ ] Colored paper
- [ ] Poster board
- [ ] Water
- [ ] Scissors

Activity 79: The Extraordinary Journey
- [ ] Legal-size piece of white paper
- [ ] Colored pencils or markers
- [ ] Straight-edge ruler
- [ ] Pencil

Activity 80: How Much Blood?
- [ ] Three index cards
- [ ] Pencil
- [ ] Five 1-liter empty bottles
- [ ] Water
- [ ] Red food coloring*
- [ ] Funnel

Activity 81: Simple Fake Blood
- [ ] Green food coloring*
- [ ] Red food coloring*
- [ ] ⅓ cup of warm water
- [ ] ⅔ cup of corn syrup
- [ ] Five tablespoons of cornstarch
- [ ] One tablespoon of cocoa powder
- [ ] Mixing bowl
- [ ] Mixing spoon

Activity 82: Easy Blood Model
- [ ] Yellow colored gelatin ½ cup + 1 tbsp.
- [ ] Five small red candies
- [ ] 50 white sprinkles (The type you put on top of cupcakes.)
- [ ] One mini marshmallow
- [ ] Water
- [ ] Pot
- [ ] 2-cup glass measuring cup

Activity 83: Another Easy Blood
- [ ] Red hots
- [ ] White jelly beans
- [ ] Sprinkles
- [ ] Karo syrup
- [ ] Mason jar

Activity 84: Multi-Sensory Hands-on Blood
- [ ] Large plastic bin or large bucket
- [ ] Red water beads (These can be found at a craft store in the aisle for floral arrangements.)
- [ ] Ping pong balls, 10–15 (You can add more. Add as many as you like.)
- [ ] Red craft foam
- [ ] Water

Activity 88: A Slimy Situation
- [ ] A living earthworm
- [ ] Petri dish or small container
- [ ] Medicine dropper
- [ ] Small glass of water
- [ ] Moistened paper towel
- [ ] Magnifying glass
- [ ] Stopwatch or clock with a second hand

Activity 89: A Road Block
- [ ] Clear rubber tubing or empty toilet paper rolls
- [ ] Scissors
- [ ] Water
- [ ] Markers
- [ ] Cotton or clay
- [ ] Paper

*Please be careful with items that can stain surfaces or clothing.

Activity 90: What's for Lunch?
- ☐ Pizza dough (You can purchase or make your own following any recipe you would like.)
- ☐ Two tablespoons of olive oil
- ☐ Marinara or pizza sauce
- ☐ Sliced pepperoni
- ☐ Mozzarella or grated parmesan cheese
- ☐ Cookie sheet
- ☐ Non-stick cooking spray

Activity 91: The Highways of the Body
- ☐ 5–6 ft. large butcher paper or white bulletin paper
- ☐ Markers
- ☐ Glue
- ☐ Scissors
- ☐ Red yarn
- ☐ Blue yarn
- ☐ White yarn
- ☐ Red construction paper
- ☐ Yellow construction paper

Activity 92: Feeling Under Pressure
- ☐ Squeeze bottle
- ☐ Meter or yard stick
- ☐ Water
- ☐ Red food coloring (optional)*

Activity 93: Are You As Strong As Your Heart?
- ☐ Tennis ball
- ☐ Stopwatch or clock with second hand
- ☐ Partner

Activity 94: Listen to My Heart
- ☐ Two funnels
- ☐ A large balloon
- ☐ Rubber tubing
- ☐ Scissors
- ☐ Drinking straw
- ☐ Modeling clay

Activity 95: The Heart and Lungs Team Up
- ☐ Two sandwich bags
- ☐ Three bag ties
- ☐ Small Styrofoam™ ball or balloon
- ☐ Red modeling clay

Activity 96: Open Heart Surgery
- ☐ Sheep or cow heart
- ☐ Butcher knife (adult supervision)
- ☐ Probe (optional)
- ☐ Meat tray or hard surface like a tray
- ☐ Gloves
- ☐ Sandwich baggie

Activity 97: Take a Stroll Through the Heart
- ☐ Burlap drop cloth or old sheet
- ☐ Fabric markers
- ☐ Fabric paint: red, black, blue, pink*
- ☐ Paint brush
- ☐ Plastic tablecloth
- ☐ Fabric pencil
- ☐ Yard stick
- ☐ Bean bag or sock stuffed with beans that is tied at the top

Activity 98: You Are What You Eat: A Heart Healthy Diet
- ☐ Old magazines
- ☐ Grocery store ads
- ☐ Assortment of foods
- ☐ Food nutrition labels

*Please be careful with items that can stain surfaces or clothing.

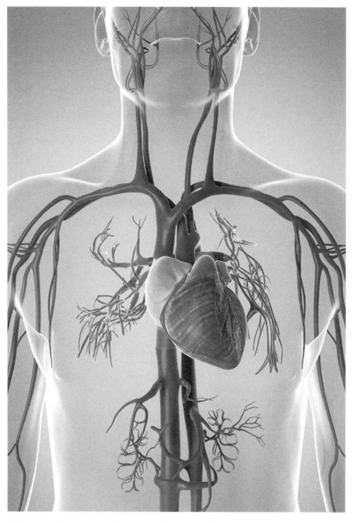

The Complex Circulatory System

Activities & Worksheets
for Use with
*Elementary Anatomy: Nervous,
Respiratory, and Circulatory Systems*

Page left intentionally blank.

| The Complex Circulatory System | After Pages 146–147 | Day 140 | Activity 75 | Name |

The Complex Circulatory System Flash Cards

Carefully cut the vocabulary cards along the dashed lines. Cards are used in multiple activities, so please store in an envelope or secure with a rubber band.

Anemia	Antibodies
Antigens	Aorta
Arteries	Arterioles
Atrium	Auscultate
Bradycardia	Bone Marrow

A blood protein that is made to attack a specific invader, like bacteria or viruses. They set off a cascade of events to assist the body in a stronger defense.	A problem with the blood in which oxygen delivered to the organs and tissues is decreased. It can be a symptom of many different diseases.
The largest artery in the body that originates from the left ventricle and sends oxygenated blood to the body.	A foreign substance, like bacteria or virus, which triggers an immune response and causes antibodies to spring into action.
Small vessels that carry oxygenated blood that connects to capillaries.	Vessels that carry oxygenated blood from the heart to the body.
To listen; to listen to the sounds of the body.	The upper chambers of the heart in which blood enters the heart. There are two atriums, the right and left atriums.
The soft spongy material in the middle of bones.	A slow heartbeat which is typically less than 60 beats per minute.

Buffy Coat	Capillary
Centrifuge	Closed Circulatory System
Chordae Tendineae	Coronary
Diastole	Electrocardiogram
Epicardium	Endocardium
Fetus	Erythrocytes

Smallest arterial blood vessel connects the arterioles with the venules.	When a blood sample is centrifuged (spun in a test tube in a machine) the blood separates into its parts. It is the middle layer, a white-colored fluid. It is composed mainly of white blood cells.
A blood system composed of vessels of different sizes which encloses the blood at all times. The blood is pumped by the heart and does not fill body cavities.	To spin around; a machine used to spin test tubes of blood at high speeds in order to cause the parts of blood to separate out.
The blood vessels that line the outside of the heart.	Fibrous strings that connect to the edges of the heart valves. They keep the valves from inverting. Also known as the "heart strings."
A machine that records a graph of the heart's electrical activity.	The phase in the heart cycle of beating in which the heart chambers relax and blood returns to the heart.
The inner muscle layer of the heart; the muscle that lines the inside of the heart.	The outer muscle layer of the heart, which lies under the pericardial sac.
A red blood cell that contains hemoglobin and transports oxygen.	An unborn baby.

Hematophagic	Hematopoiesis
Hemophilia	Hemostasis
Malaria	Myocardial Infarction
Myocardium	Open Circulatory System
Pericardial Sac	Platelets
Prothrombin	Red Blood Cells

The formation of blood cells. In the fetus, it takes place at sites including the liver, spleen, and thymus. From birth throughout the rest of life it is mainly in the bone marrow.	The act of an animal or insect like a mosquito drinking blood.
Stopping the escape of blood by natural means (either clot formation or vessel spasm).	Any of several X-linked genetic disorders, symptomatic chiefly in males, in which excessive bleeding occurs owing to the absence or abnormality of a clotting factor in the blood.
A heart attack.	A disease transmitted by mosquitos in which a parasite infects the red blood cells. A disease that can be deadly.
A blood system in which the blood is pumped by the heart and fills the body cavities. The blood does not stay within the vessels.	The middle and thickest layer of the heart wall muscle.
Small cells in the blood which are important in hemostasis, forming blood clots.	Fibrous, double-layered sac that surrounds the heart. It is filled with a lubricant that allows the heart to move without friction.
Cells in the blood that contain hemoglobin, an iron. It carries oxygen.	A clotting factor, made in the liver, which is in the blood. It is activated to thrombin for clot formation.

Sickle Cell Anemia	Sinoatrial Node
Stem Cell	Stethoscope
Syncope	Systole
Tachycardia	Tricuspid Valve
Universal Donor	Universal Receiver
Valves	Veins

A mass of muscle tissue on the top of the right atrium which is the electrical pacemaker of the heart.	A blood disease that is inherited in which the red blood cells become misshapen to a sickle-like appearance. Causes long-term problems.
A medical device used to listen and magnify the sounds heard in the body.	A cell that has the ability to differentiate to other specialized cells.
The phase in the heart cycle of beating in which the heart chambers contract to expel blood out of the heart.	To lose consciousness; pass out.
The heart valve between the right atrium and right ventricle.	A fast heart beat which typically is over 100 beats per minute.
A person who has type AB blood.	A person who has type O blood.
Blood vessels in the body that carry deoxygenated blood. They transport blood to the heart.	The doors between the chambers of the heart that prevent blood from flowing backwards.

Ventricles	Venules

White Blood Cells	

Small blood vessels that carry deoxygenated blood toward the heart. They connect the capillaries to the veins.	The lower chambers of the heart.
	Blood cells that are part of the immune system; they fight invaders that attack the body.

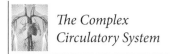

The Complex Circulatory System | After Pages 148–149 | Day 141 | Activity 76 | Name

Cardiac Cobble Stones

Make several copies of the jumping cardiac cobble stones, as seen below and the next page. Cut the stones out and write out one of the selected words from the vocabulary list. Place one word on each of the copied stones. Lay the "stones" out on the floor. Call out a vocabulary word and have the student jump on the designated word. For an extra level of difficulty read the definition of the word and have the student jump to the designated word.

| The Complex Circulatory System | After Pages 150–151 | Day 142 | Worksheet 54 | Name |

Copywork

A glad heart makes a happy face;

   a broken heart crushes the spirit.

A wise person is hungry for knowledge,

   while the fool feeds on trash.

               Proverbs 15:13-14 (NLT)

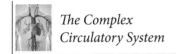

Copywork

A glad heart makes a happy face;

a broken heart crushes the spirit.

A wise person is hungry for knowledge,

while the fool feeds on trash.

Proverbs 15:13-14 (NLT)

The Complex Circulatory System | After Pages 152–154 | Day 143 | Worksheet 56 | Name

Heart Matters and Its Meaning

Fill in the blanks as you read the text in the section "Biblical References to the Heart."

The Bible mentions the heart over 900 times. Why do you think it is that the heart gains so much attention in the Bible?

What are the seven themes that the Bible ties to the heart?

1. The heart _____, _____, _____, is wise _____ and _____.

2. The heart _____, _____, and _____.

   What does it mean to envy?

3. The heart is very _____.

4. The heart is _____. It _____. It _____ things good and bad.

5. The heart can be _____, _____, and _____.

6. The heart can be _____, _____, and _____.

7. The heart can be _____ and store _____.

Look back at all the verbs you filled in the previous blanks. List these active qualities of the heart in the chart below. List the verbs in one of the two columns. Write any good qualities under the "thumbs up" column. Write any bad qualities under the "thumbs down" column. For any quality that has the potential of being good or bad, write those across the middle divider of the columns.

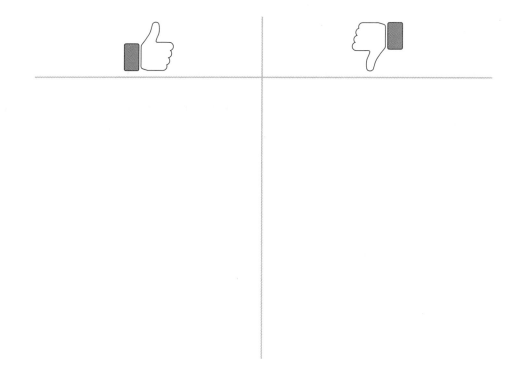

| The Complex Circulatory System | After Pages 155–157 | Day 144 | Worksheet 57 | Name |

The Circulatory System Report

Pick from one of the topics below. Give an oral or written report on the subject. Information is available on pages 17–22 of *The Complex Circulatory System*.

Oral Report

- Research a famous person who contributed to the advancement of our knowledge of the circulatory system.

- Give an oral report. (Follow the rubric for oral reports in the assessment section of this manual.)

    OR

- Assume creatively the identity of the person you researched. Dress in period clothing and have someone conduct a television interview in which you are showcased for your contribution to science.

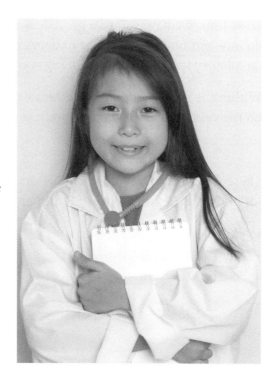

Written Report

Write a report about a famous person/scientist that contributed to the field of cardiology. The length of the paper should be determined by your instructor. Follow the rubric for written reports in the assessment section of this manual.

Ideas of Famous Persons

William Harvey	Medieval Barber	Leonardo da Vinci
Andreas Vesalius	Helen Taussig	Daniel Hale Williams
Charles Richard Drew	Karl Landsteiner	Aristotle
Willem Einthoven	Michael DeBakey	Alfred Blalock
Christian Barnard	Adrian Kantrowitz	Hippocrates
Clara Barton	Getrude Belle Elion	

Early Ideas on the Circulatory System

Over time, we have gained a great deal of knowledge on how our bodies work. Our understanding of the circulatory system in the beginning was very limited.

Within the person outline below, draw the early beliefs on the veins and arteries in the body.

These misconceptions of the circulatory system existed for hundreds of years. Why did this happen?

What was one misconception about the circulatory system?

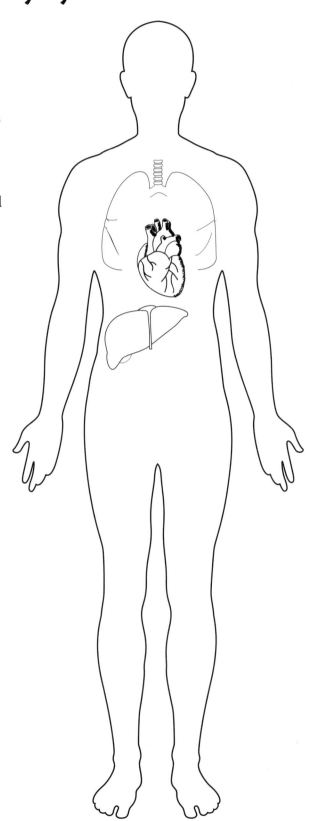

 | The Complex Circulatory System | After Pages 158–161 | Day 146 | Activity 77 | Name

Timeline Shuffle

Cut out the following images and paste them in the appropriate order on the timeline.

Hippocrates

Herophilus

Claudius Galen

Barbers as surgeons

Leonardo da Vinci

Michael Servetus

Page left intentionally blank.

 Andreas Vesalius	 William Harvey	 Richard Lower
 Marcello Malpighi	 James Blundell	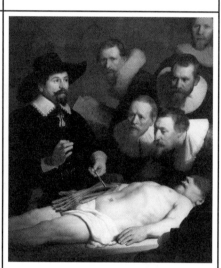 Professionals can use donated bodies for study and dissection.
 Daniel Hale Williams	 Ernest Henry Starling	 Willem Einthoven

Page left intentionally blank.

Karl Landsteiner

Jules Bordet

Werner Forssmann

Charles Richard Drew

Helen Brooke Taussig

Michael DeBakey

Christiaan N. Barnard

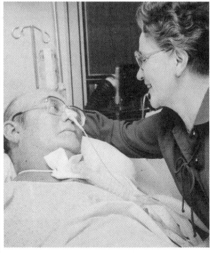
Barney Clark

Dr. Drew Gaffney

Page left intentionally blank.

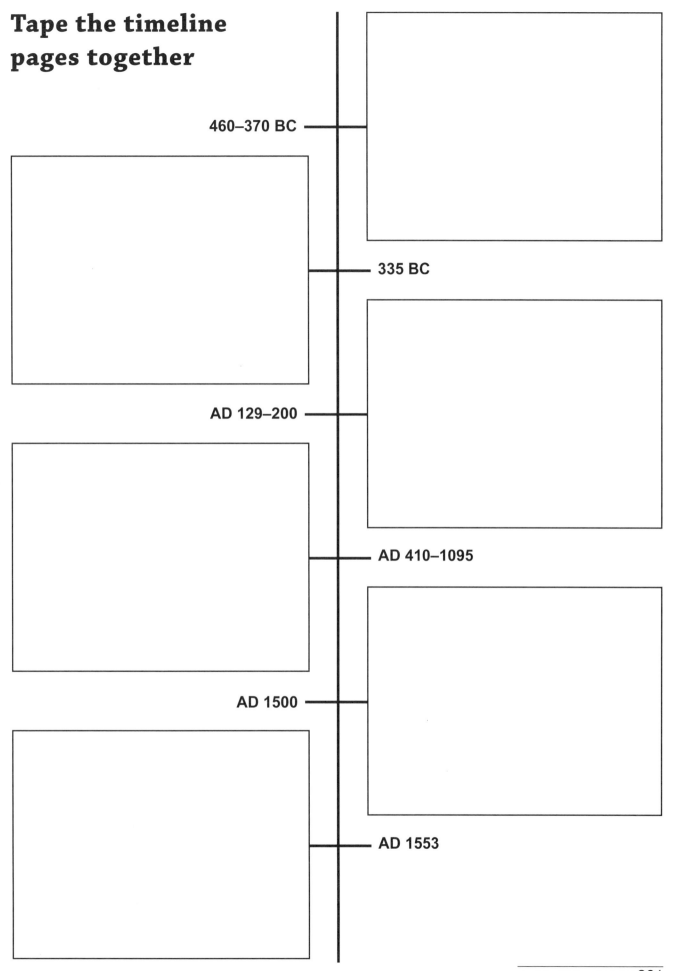

Page left intentionally blank.

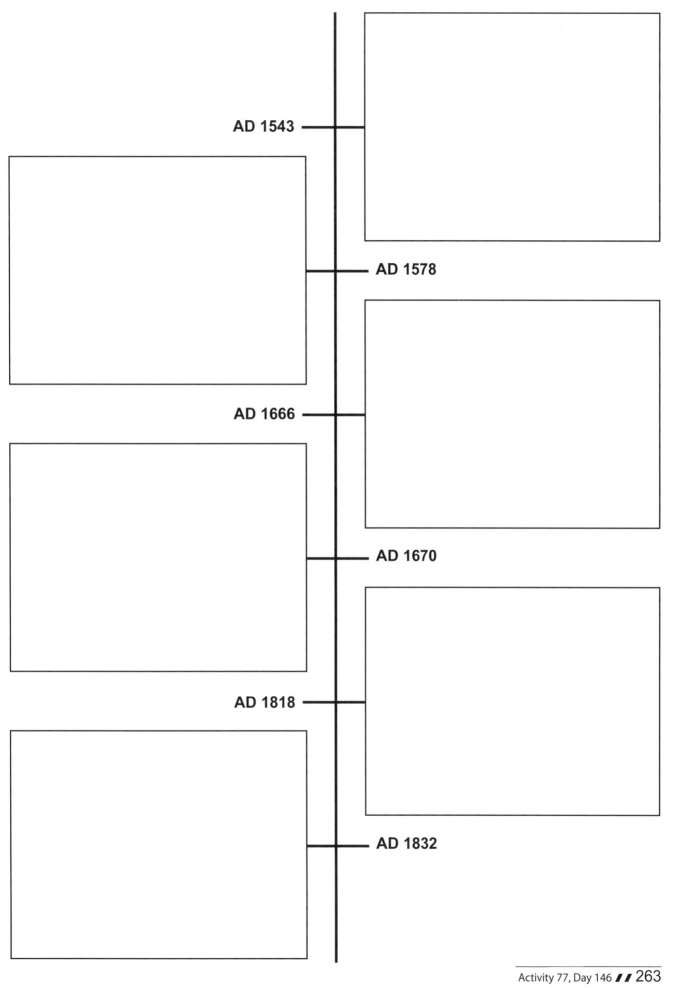

Page left intentionally blank.

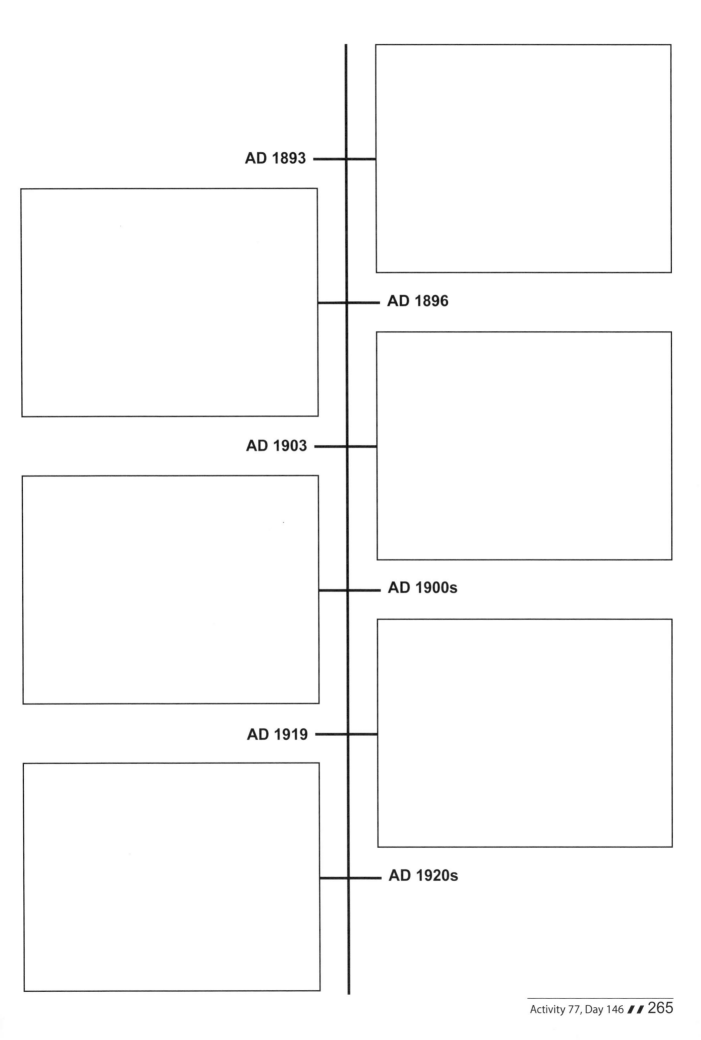

Page left intentionally blank.

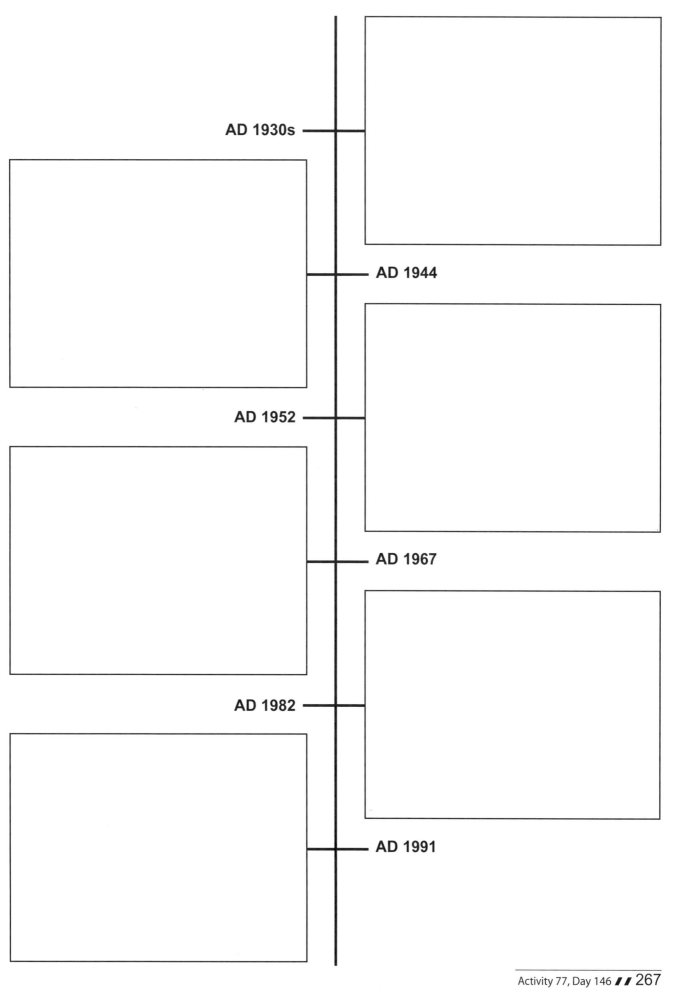

Page left intentionally blank.

| The Complex Circulatory System | After Pages 162–166 | Day 147 | Activity 78 | Name |

Blood Splatter Art

Materials:

- Paints
- Paint brush
- Toothbrush
- Glitter
- Newspaper
- School glue
- Colored paper
- Poster board
- Water
- Scissors

Instructions:

1. Prepare your work area by spreading out newspaper to product the surfaces on which you will be working.

2. Place your paper on top on the newspaper.

3. Make sure your paint is runny. If needed, add water to the paints so that they are runny like water.

4. Take some paint on your paintbrush and flick your wrist to splatter the colors on the paper. Do this with as many colors as you like. (Red and blue would be great choices if you like. Be creative.)

5. You can also add paint to your toothbrush and tap the paper.

6. Take your glue and squeeze circles out on the paper. Make the circles throughout the painting.

7. While the glue is still wet, sprinkle glitter onto the glue. Make sure you do not go too heavy.

8. Allow the picture to dry overnight.

9. The following day, rim the edges of the paper in whatever pattern you would like.

10. Glue your picture on the cardboard.

11. Magnificent…a picture fit for framing and hanging in the Chicago Museum of Art!

| The Complex Circulatory System | After Pages 162–166 | Day 147 | Worksheet 59 | Name |

Help Wanted Ad

You are the human resources guru for the company, The Human Body, Inc. Your job is to write a help wanted ad in the local paper to fill the open position of "Sanguine Fluid Executive Director." Give a complete description of the job requirements, hours, expectations, holidays, and salary. To aid in your brain storming session with your executive crew, list all the functions of the blood below.

Functions of Blood:

1. _____
   _____
   _____

2. _____
   _____
   _____

3. _____
   _____
   _____

4. _____
   _____
   _____

5. _____
   _____
   _____

6. _____
   _____
   _____

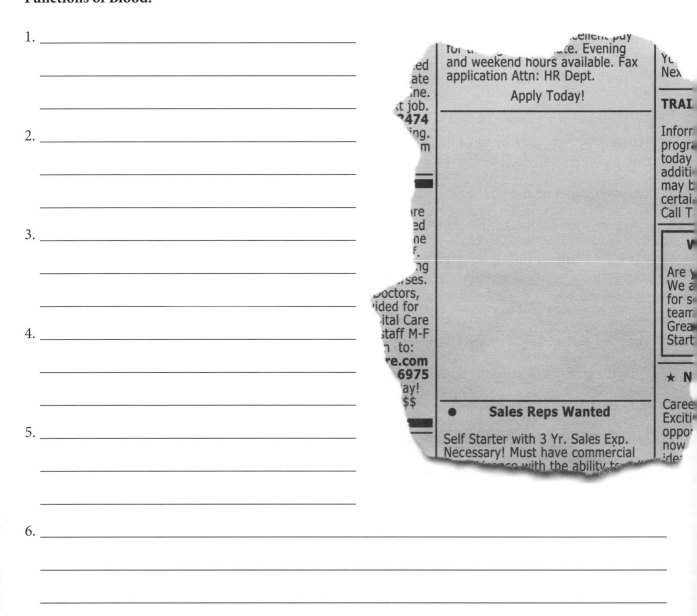

| The Complex Circulatory System | After Pages 162–166 | Day 148 | Activity 79 | Name |

The Extraordinary Journey

Materials:

- Legal-size piece of white paper
- Colored pencils or markers
- Straight-edge ruler
- Pencil

You are living the life of a platelet. You are floating down the "lazy" river of life, when suddenly you are called into action. Write a comic strip depicting your daily life. Then, show what happens when you join in the party of a clot formation. (A template is included on the back of this page. You may choose to draw your own.)

Ideas for formatting:

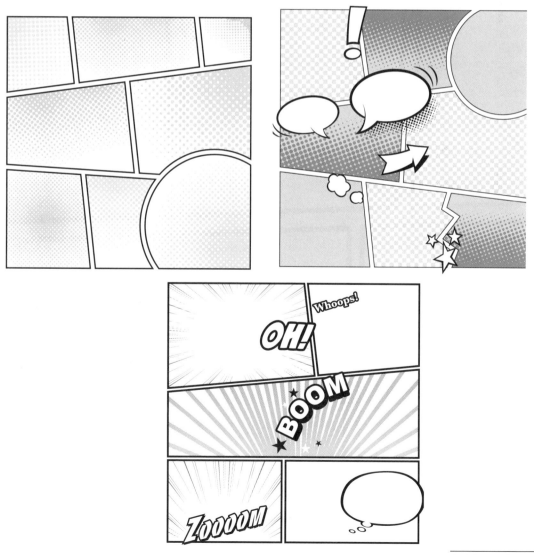

| The Complex Circulatory System | After Pages 162–166 | Day 148 | Worksheet 60 | Name |

Parts of the Blood

Below is a picture of a blood sample that has been centrifuged.

What does centrifuge mean?

_____

_____

_____

Label the layers of the blood sample on the left side. Draw an arrow pointing to each of the areas. On the right side, label what each layer of the blood in the tube is composed of.

Parts of the Blood Sample   Composition of the layer

1. _____   _____

   _____   _____

   _____   _____

   _____   _____

2. _____   _____

   _____   _____

   _____   _____

   _____   _____

3. _____   _____

   _____   _____

   _____   _____

   _____   _____

Put Some Color in It

Color three red blood cells red.

Color four platelets yellow.

Color five white blood cells green.

Color two Band-Aids® blue.

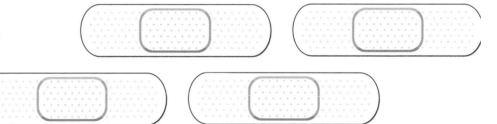

Color the serum yellow.

Color the white blood cells white.

Color the red blood cells red.

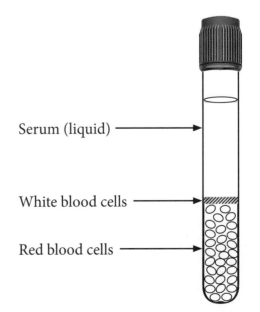

Serum (liquid)

White blood cells

Red blood cells

| The Complex Circulatory System | After Pages 162–166 | Day 148 | Worksheet 62 | Name |

Blood Smear

Below is a blood smear.

Circle a red blood cell in green.

Circle a white blood cell in blue.

Circle a platelet in brown.

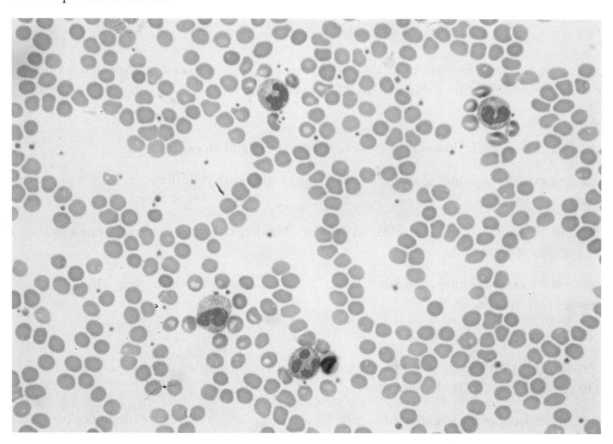

What is the function of each of these cells?

1. Red Blood Cell _____

_____

2. White Blood Cell _____

_____

3. Platelet _____

_____

| The Complex Circulatory System | After Pages 167–169 | Day 149 | Activity 80 | Name |

How Much Blood?

Materials:

- 3 index cards
- Pencil
- 5 1-liter empty bottles
- Water
- Red food coloring
- Funnel

As you grow up, your body increases in its total blood volume. A baby has less blood than an adult does. Let's design a visual representation of those blood differences.

Instructions:

1. Fold the index cards in half. This will allow the index cards to stand up easily.

2. On index card #1, draw a picture of a baby. Write "Baby" underneath your drawing. Draw a picture of one bottle next to the baby.

3. On index card #2, draw a picture of a child. Write "Child" underneath your drawing. Draw a picture of three bottles next to the child.

4. On index card #3, draw a picture of an adult. Write "Adult" underneath your drawing. Draw a picture of five bottles next to the adult.

5. Fill the five 1-liter bottles with water.

6. Add approximately 10–15 drops of red food coloring to each of the bottles. Add as many drops as needed to achieve the desired color intensity.

7. Place the cards in front of the five colored bottles. Now you can actually visualize the amount of blood volume that you and your family members have flowing inside of them.

Simple Fake Blood

CAUTION: Remember, food coloring can stain your skin and clothing — make sure you have permission to make and have fun with this recipe.

Materials:

- Green food coloring
- Red food coloring
- ⅓ cup of warm water
- ⅔ cup of corn syrup
- 5 tablespoons of cornstarch
- 1 tablespoon of cocoa powder
- Mixing bowl
- Mixing spoon

Instructions:

1. In the bowl, mix the corn syrup, water, and cornstarch.

2. Once the items are thoroughly mixed, add the red and green food coloring. Mix.

3. Add the cocoa powder. To obtain the desired color of the blood, feel free to add more cocoa powder or food coloring. Mix thoroughly.

4. Drizzle it onto your hand, arm, or face.* Enjoy.

What Color is Your Blood?

Solve clues 1–7 and fill in the crossword puzzle below.

Across:

1. Vessel that carries oxygenated blood
2. Largest vessel in the body
3. Person who gives blood

Down:

4. The red liquid substance that runs through your vessels

5.

6. Vessel that carries deoxygenated blood
7. The _____ coat is the middle section of a centrifuged blood sample.

| The Complex Circulatory System | After Pages 170–171 | Day 151 | Activity 82 | Name |

Easy Blood Model*
*REQUIRES ADULT SUPERVISION.

Materials:

- ½ cup + 1 tbsp. yellow-colored gelatin
- 5 small red candies
- 50 white sprinkles (The type you put on top of cupcakes.)
- 1 mini marshmallow
- Water
- Pot
- 2-cup glass measuring cup

Instructions:

1. Ask for adult assistance with this first step. Place water in pan and boil. Take the gelatin powder and mix with ⅓ cup of boiling water.

2. Stir until all of the gelatin powder is completely dissolved.

3. Add ½ a cup of cold water to the dissolved gelatin powder. Stir well.

4. Put the gelatin in the refrigerator for up to an hour. This will allow the mixture to thicken without completely solidifying.

5. Once the mixture is thickened, but NOT solidified, stir in the small red candies. What do these represent?

6. Drop in the white sprinkles and marshmallow. Stir. What do the white sprinkles and marshmallow represent?

7. There you have it. Your easy blood model!

***NOTE:** This recipe can be done with plain gelatin, food coloring, and various fruits. While they would not be as reflective of the actual shape of the blood components, it could still serve as a simple model while students can review the book for actual images of the components of blood. If using any kind of food coloring, please remember it can stain both clothing and skin.

Another Easy Blood Model

Materials:

- Red hots
- White jelly beans
- Sprinkles
- Karo or other corn syrup
- Mason jar

Instructions:

1. Take a mason jar and fill with a mixture of the candies listed above.

2. Add corn syrup to the jar. Add enough to cover all the candy.

3. Gently mix the candy and syrup together.

There you have it . . . your own easy blood model.

Put a Plug in It

Hemostasis is the process in which the body stops you from bleeding in the event of an injury. God has designed the process to occur in a series of steps. Each step depends on the previous step to happen for the clotting process to be effective.

Robert has a cut. He is bleeding. You must help his body stop the bleeding as soon as possible. Below you will see a series of pictures. Color the pictures. Cut them out and place the pictures in the order of the clotting process so that Robert will be able to get back to having fun.

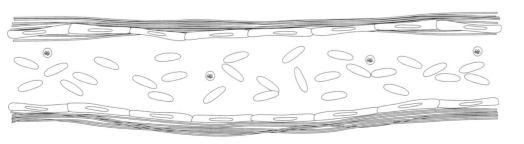

Multi-Sensory Hands-on Blood

This is a perfect activity for the sensory learner.

Materials:

- Large plastic bin or large bucket
- Red water beads (These can be found at a craft store in the aisle for floral arrangements.)
- 10 to 15 ping pong balls (You can add more. Add as many as you like.)
- Red craft foam
- Water

Instructions:

1. Place all the water beads in the bucket. Follow the directions to hydrate the beads by adding water. Allow the beads to soak up the water overnight. Typically, it takes 8 to 10 hours for the beads to soak up the water.

2. Cut up the red craft foam into small bits and pieces. These small pieces will represent the platelets in your blood.

3. The following day, place the pieces of red craft foam and ping pong balls into the water beads.

4. Mix all the items in the bucket. Squish and play.

| The Complex Circulatory System | After Pages 172–173 | Day 152 | Activity 85 | Name |

Blood Transfusion Simulation

Materials:

- 8 small paper drinking cups
- Yellow food coloring
- Blue food coloring
- Permanent marker
- Water

You are an emergency room doctor. You have been called into the trauma room. You are awaiting the ambulance's arrival with a critically injured patient. The Emergency Technician has radioed you. You have been informed that the patient has lost a great deal of blood from his injury. You ready the room. You call the hospital blood bank to prepare blood for your patient. You know that blood is classified into four basic groups: A, B, AB and O. These blood groups are dependent on the type of antibodies, a specific type of protein, found in the blood. O blood is the universal donor because it possesses no antibodies. Your job is to determine which blood groups will safely mix with the patient's blood in a transfusion.

Instructions:

1. Take the eight cups. With your permanent marker label two cups each with A, B, AB and O. On four of the cups that are labeled "A," "B," "AB," and "O" write "donor" below the blood type. Label the other four remaining cups "Recipient" under the blood types.

2. Fill all the cups halfway with water.

3. In cups A, add 2 drops of yellow food coloring.

4. In cups B, add 2 drops of blue food coloring.

5. In cups AB, add 2 drops of blue food coloring and 2 drops of yellow food coloring.

Now you are ready to test your patients.

6. Draw up a small amount of fluid from the "A" donor blood cup and place 2 drops in each of the recipient's cups. If the recipient is able to receive the "A" donor blood safely, then the color will not change. The patient is safe to receive a blood transfusion with "A" blood. If the color changes, the transfusion is not safe and could be deadly. In an actual transfusion, reactions occur when an inappropriate blood type is transfused to the recipient. The blood experiences an agglutination reaction. Do you remember what that is? Research the word "agglutination."

7. Fill out your results below in the Data Chart. Give a "thumbs up" if the transfusion would be safe. If the transfusion would be unsafe, give a "thumbs down."

Donor Blood Type	Recipient Blood Type			
	A	B	AB	O
A				
B				
AB				
O				

8. Rinse out all the recipient cups from the prior trial. Refill and repeat #3–5. Now place 2 drops of "B" blood in each of the recipient cups. Record results.

9. Rinse. Refill and repeat #3–5. Now place "AB" blood in each of the recipient cups. Record results.

10. Rinse. Refill and repeat #3–5. Now place "O" blood in each of the recipient cups. Record results.

Exploratory Questions:

1. Which blood types can "A" recipient safely accept? _____

2. Which blood types can "B" recipient safely accept? _____

3. Which blood types can "AB" recipient safely accept? _____

4. Which blood types can "O" recipient safely accept? _____

| The Complex Circulatory System | After Pages 172–173 | Day 152 | Activity 86 | Name |

Research Blood Transfusions

1. What parts of blood can be transfused into a recipient?

    A. Platelets

    B. Plasma

    C. Red Blood Cells

    D. All of the Above

2. What is the minimum you should weigh in order to donate blood?

    A. 90 pounds

    B. 100 pounds

    C. 110 pounds

    D. 115 pounds

3. How often can a donor give blood?

    A. All the time

    B. Once a year

    C. Every 2 months

    D. Every 6 months

4. How much blood can a donor give at one time?

    A. 1 pint

    B. 2 pints

    C. 3 pints

    D. 1 gallon

5. What organization regulates blood donation?

    A. American Lung Association

    B. Federal Food and Drug Administration

    C. American Red Cross

    D. American Bodily Fluids Association

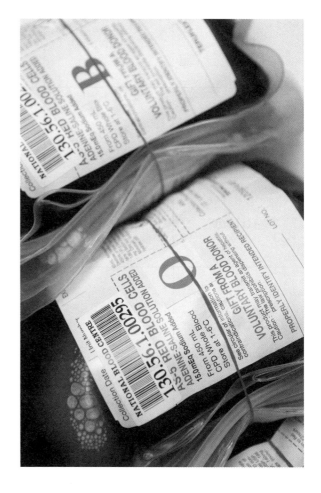

Pick from one of the following research questions below and write a one-page paper on the subject. (Length of research paper will vary based on the instructor's preferences and the level of the student.)

1. Research two diseases that can be transmitted via a blood transfusion.

2. What symptoms may a patient experience with a blood transfusion reaction?

3. How is blood collected for a blood transfusion?

| | The Complex Circulatory System | After Pages 172–173 | Day 152 | Worksheet 65 | Name |

Color Your Heart

Help the blood pump through the heart. Color the boxes with the word **Heart** in red.

Blood	Atrium	Valve	Systole	Heart
Heart	Valve	Systole	Blood	Valve
Blood	Blood	Heart	Atrium	Systole
Atrium	Heart	Blood	Systole	Heart
Heart	Systole	Valve	Heart	Systole
Valve	Blood	Systole	Atrium	Blood
Valve	Systole	Heart	Atrium	Heart
Systole	Blood	Systole	Systole	Valve

| The Complex Circulatory System | After Pages 172–173 | Day 152 | Worksheet 66 | Name |

The Blood Report

Pick from one of the topics below. Give an oral or written report on the subject.

Oral Report

Research a blood disease. Give an oral report. (Follow the rubric for oral reports in the assessment section of this manual.)

Written Report

Write a report about a blood disease. The length of the paper should be determined by your instructor. Follow the rubric for written reports in the assessment section of this manual.

Ideas of Different Blood Diseases

Anemia	Hemophilia	Burkitt's Lymphoma
Sickle-Cell Anemia	Malaria	Deep Vein Thrombus
Leukemia	Fanconi Anemia	Hemolytic-Uremic Syndrome
Thalassemia	Hodgkin's Lymphoma	Protein C Deficiency
Toxoplasmosis		

You Are Not My Type

Blood types were discovered by Karl Landsteiner. His discovery solved the puzzling problems of blood transfusions. Below you will see four donors who have arrived at the local blood drive to donate their blood. Utilizing four different colored pencils, draw colored lines from each donor to the recipients that can receive their blood.

Blood Donor Types

O   A   B   AB

Recipient Blood Types

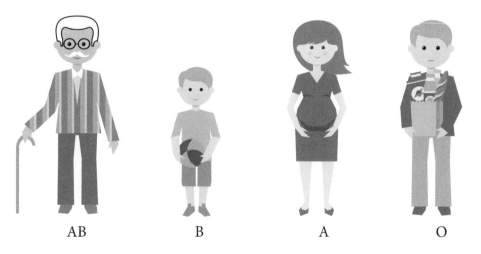

AB   B   A   O

| The Complex Circulatory System | After Pages 174-175 | Day 153 | Activity 87 | Name |

Match a Blood Sucker

NOTE: Cut out these cards after gluing the page to cardstock paper, or cut out and glue to index cards. Use as a simple matching/memory exercise game.

1. Mix up the cards.
2. Lay them in rows, face down.
3. Turn over any two cards.
4. If the two cards match, keep them.
5. If they do not match, turn them back over.
6. Remember what was on each card and where it was.
7. Watch and remember during the other player's turn.
8. The game is over when all the cards have been matched.
9. The player with the most matches wins.

Vampire Bat

Vampire Bat

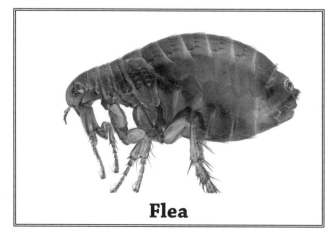
Flea

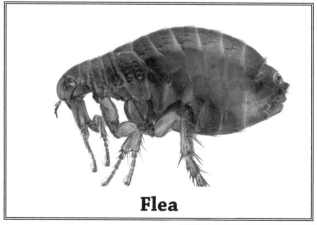
Flea

Page left intentionally blank.

Bed Bug

Bed Bug

Leech

Leech

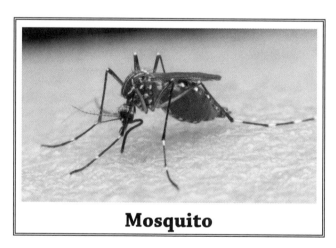
Mosquito

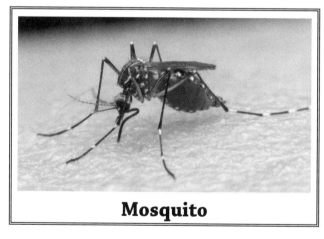
Mosquito

Lamprey

Lamprey

Page left intentionally blank.

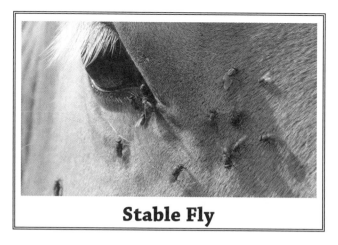
Stable Fly

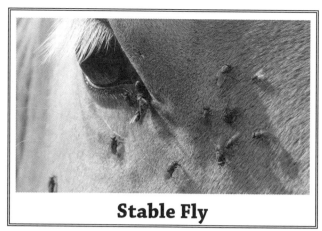
Stable Fly

Head Louse

Head Louse

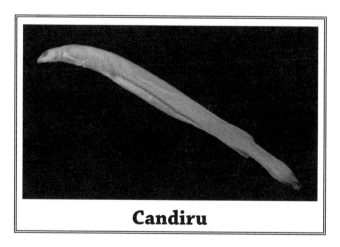
Candiru

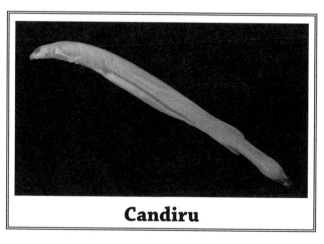
Candiru

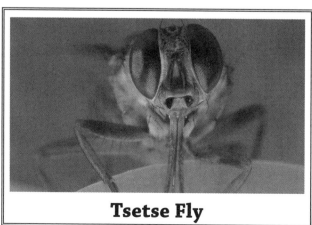
Tsetse Fly

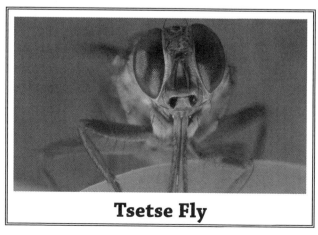
Tsetse Fly

Page left intentionally blank.

Botfly Larvae

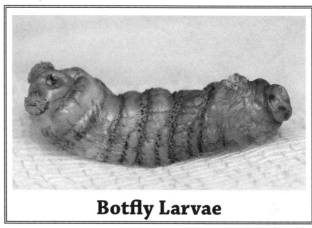
Botfly Larvae

Deer Tick

Deer Tick

Vampire Squid

Vampire Squid

Assassin Bug

Assassin Bug

Page left intentionally blank.

| The Complex Circulatory System | After Pages 174–175 | Day 153 | Worksheet 68 | Name |

A Day in the Life of a Blood Cell

Write the title of your story on the line below. Be sure to include your journey through the heart and lungs. Draw a picture in the box to accompany your story.

Title: _____

Blood Factories

Each of your blood cells has a life expectancy of about 115 days. Your body has its own manufacturing plants to replace old blood cells. Blood is manufactured in specific bones of your body. Circle the pictures of the bones in which Hematopoiesis occurs in adults.

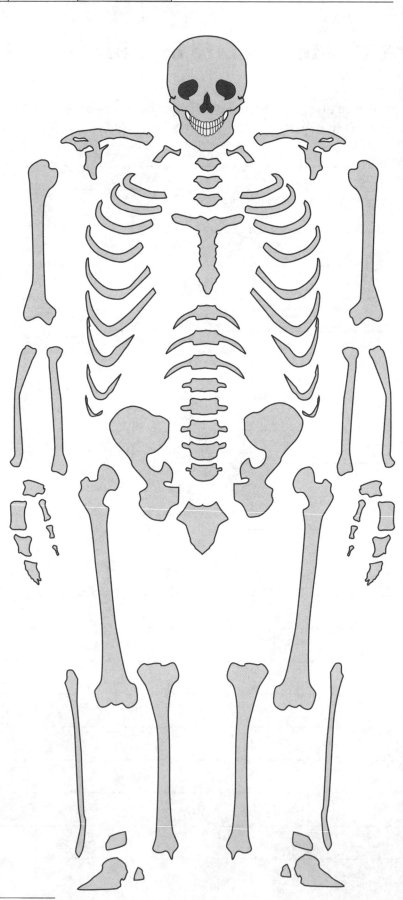

| The Complex Circulatory System | After Pages 174–175 | Day 153 | Worksheet 70 | Name |

Drawing Comparisons

The arteries and veins are the highway systems of the body that transport blood. Compare arteries with veins. How are they similar and different?

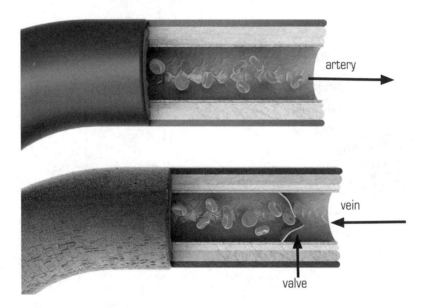

	Arteries	**Veins**
Similar		
Different		

The Highways of Blood Report

Write a one-page report on one of the following diseases listed below. Then develop a trifold brochure, educating the public on the symptoms, causes, risk factors, and healthy lifestyle changes to lower a person's risks.

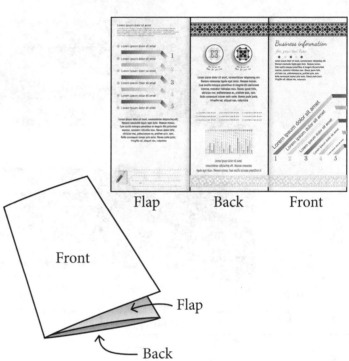

Ideas

Heart Disease

Arteriosclerosis

Thrombus

Stroke

Hypertension

Diabetes

Obesity

Effects of Smoking on the Heart

| The Complex Circulatory System | After Pages 176–177 | Day 154 | Activity 88 | Name |

A Slimy Situation

Materials:

- A living earthworm (Be gentle!)
- Petri dish or small container
- Medicine dropper
- Small glass of water
- Moistened paper towel
- Magnifying glass
- Stopwatch or clock with a second hand

Earthworms have a thing or two in common with us when it comes to the circulatory system. There are certainly differences between us. In this simple lab, you will use your power of observation to discover how we are similar and different from the earthworm.

Instructions:

1. Take your moistened paper towel and fold it large enough to cover the bottom of the container.

2. Place a living wiggly earthworm on top of the moist paper towel in your dish.

3. Be sure to use your dropper and water to help keep the little fellow moist. Add only a couple of drops occasionally. Do not add too much. You want to keep him happy and healthy so that you can let him go after you are done.

4. Hold your magnifying glass a few inches above the earthworm. Watch the surface of the earthworm's skin closely through the lens. Do you see pulsating, or an area where the skin is moving up and down? If so, you have found the pulse of the earthworm.

5. Once you have located the pulse, count how many times in a minute it pulses. Record it in your chart. Do this three times and record your data. (This process is called a trial.)

6. Find your pulse. Your pulse can be located behind the angle of your jaw. Place your index and middle fingers over the area. You may have to move your fingers around slightly to find your pulse. (Do not use your thumb. Your thumb has its own pulse.) Record your pulse for a minute. Do three times and enter the numbers into the data chart.

7. Upon completion of the measurements, calculate the average heart rate for the earthworm and yourself.

How do you calculate an average?

1. Add the results of these trials together for the earthworm and then do the same process for your pulse.

    Trial #1 + Trial #2 + Trial #3 = Total

2. Total and divide it by the three trials — this will give you the average.

    Total ÷ 3 = _____ **Average heart rate for earthworm**

    Total ÷ 3 = _____ **Average heart rate for yourself**

Your Data:

Experimental Trials	Earthworm	Fearfully, Wonderfully Made You
1		
2		
3		
Average		

Questions:

1. Why would you think there is a need to do several trials in an experiment?

   _____
   _____

2. How does an earthworm's pulse compare to yours?

   _____
   _____

3. What similarities do you have with an earthworm? (Hint: It is a type of circulatory system.)

   _____
   _____

4. Do a bit of research on the circulatory system of an earthworm. How is it different from yours?

   _____
   _____

| The Complex Circulatory System | After Pages 176–177 | Day 155 | Worksheet 72 | Name |

Blood-Sucking Critters

Let's welcome our Academy Award winners for blood sucking. Match the three critters with the following statements. Fill in the corresponding letter next to the statement description for the correct blood-loving creature.

        A. Mr. Tiberius Tick

        B. Miss Mosey Mosquito

        C. LeeWynn Leech

1. _____ I am a type of worm.
2. _____ I am used for Hirudotherapy.
3. _____ My wings flap up to 400 times per second.
4. _____ I prefer digging in and dining behind the neck and ears.
5. _____ My husband hates the taste of blood. He prefers sweet, disgusting nectar.
6. _____ I love giving gifts like Rocky Mountain Spotted Fever and Lyme Disease.
7. _____ I prefer to feed on people with Type O blood. Yummy!
8. _____ I have no time for love. I am all I need wrapped in one. I am a hermaphrodite.
9. _____ The dinner bell rings for me when I sense carbon dioxide.
10. _____ I hate garlic and lavender. Yuck.
11. _____ I secrete hirudin so my prey cannot form a blood clot.
12. _____ I live fast and furious for 14 days.
13. _____ I prefer deer…but you make a yummy snack also.
14. _____ I am considered a medical device.

Circulatory System Scramble

Unscramble the circulatory system vocabulary words on the hearts and write them on the opposite heart.

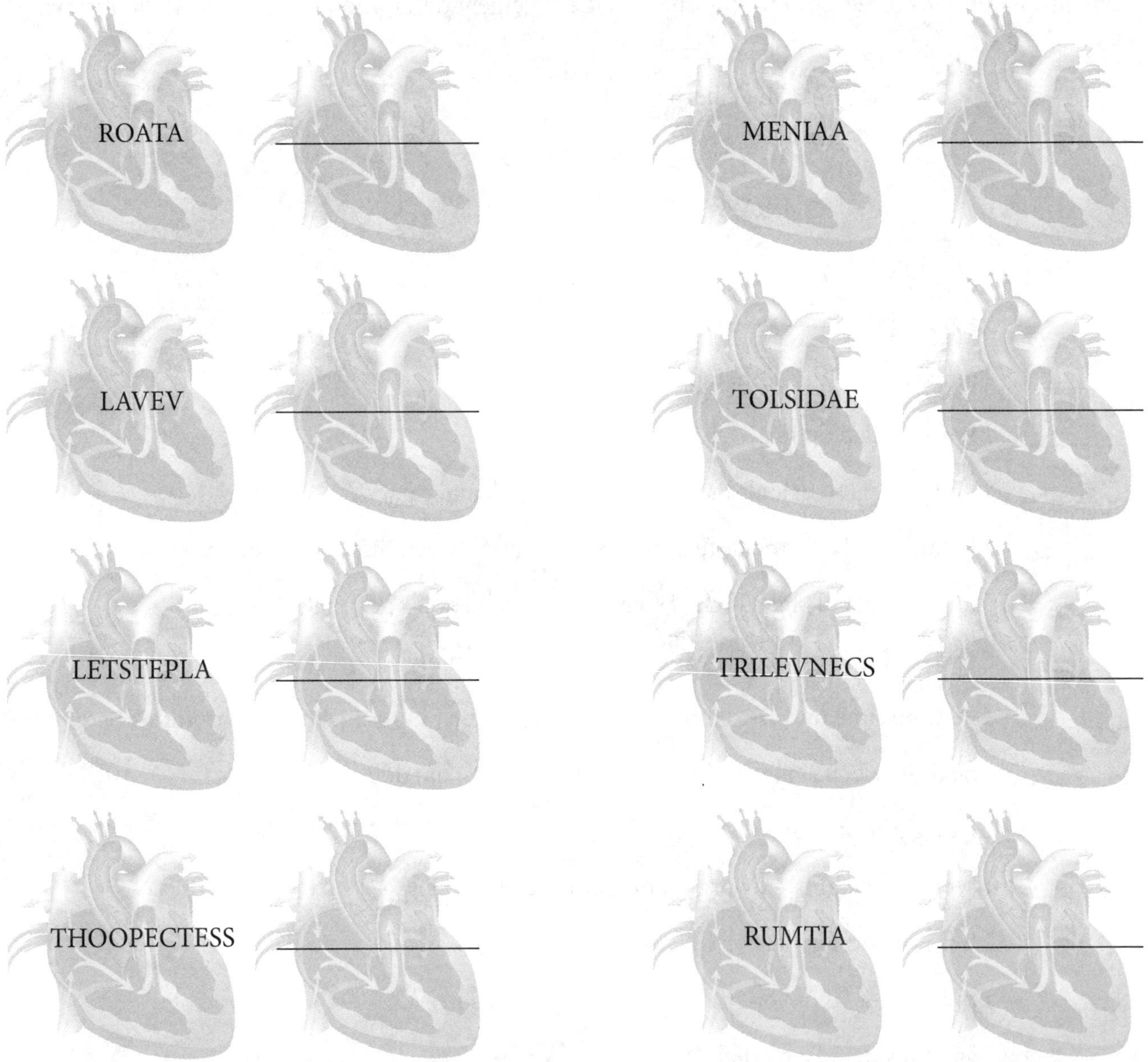

ROATA _____

MENIAA _____

LAVEV _____

TOLSIDAE _____

LETSTEPLA _____

TRILEVNECS _____

THOOPECTESS _____

RUMTIA _____

TYOLESS _____

CARDBDRAAYI _____

The Complex Circulatory System | After Pages 178–179 | Day 156 | Worksheet 74 | Name

Copywork

Jesus replied: "'Love the Lord your God with all your heart and with all your soul and with all your mind."

Matthew 22:37

Finding Your Place

There are approximately 60,000 miles of blood vessels in your body.

How many miles of blood vessels total would there be in your household if everyone's were added together?

60,000 x _____ (number of people in your family) = _____ miles

Write the number in the below boxes and then write above the place value of each of the numbers.

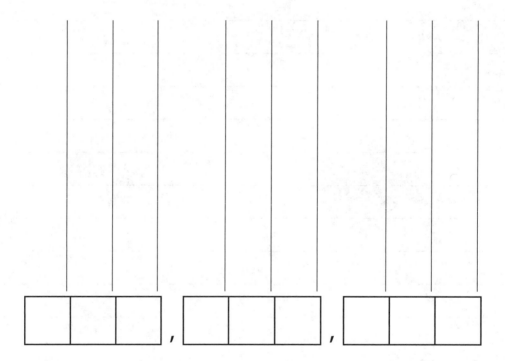

| The Complex Circulatory System | After Pages 180–181 | Day 157 | Worksheet 76 | Name |

Closed vs. Open Circulatory Systems

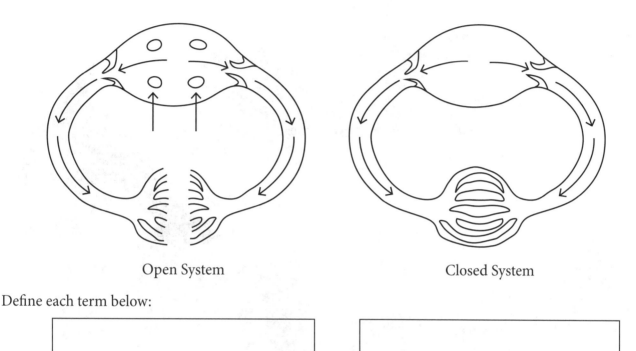

Open System   Closed System

Define each term below:

God has created in his creatures two types of circulatory systems. Each type of circulatory system serves its organism quite well. Circle the animals that have an open circulatory system with blue. Circle the animals that have a closed circulatory system with red.

Your Beating Heart

How does your heartbeat change?

I counted _____ heartbeats when I

I counted _____ heartbeats when I

I counted _____ heartbeats after a minute of

| The Complex Circulatory System | After Pages 182–183 | Day 158 | Activity 89 | Name |

A Road Block

Materials:

- Clear rubber tubing or empty toilet paper rolls
- Scissors (for cutting paper or rolls)
- Cotton or clay
- Paper

Instructions:

1. Use the picture below of a cross section and longitudinal section of an artery as your example for your mission to design and build models of a healthy artery and an unhealthy artery.*

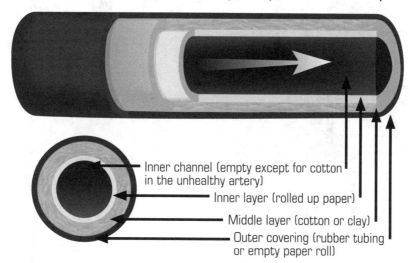

2. How is blood flow different between a healthy and unhealthy artery?

___

___

3. How is a heart attack related to an unhealthy artery?

___

___

*__Note__: You can use the clear tubing or a rolled up piece of paper (inner layer) within the empty paper roll (outer covering). Either cotton or clay could be filler (middle layer). Use cotton or clay to show unhealthy artery features like blockage, etc.

The Pathway of Your Circulating Blood

Complete the graphic organizer below. Outline the flow of blood through the entire circuit of the body.

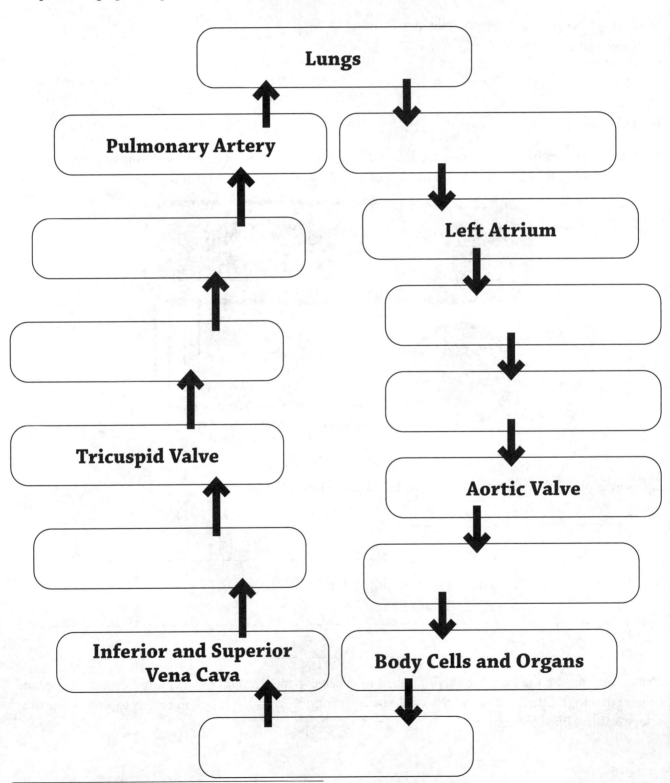

The Complex Circulatory System | After Pages 182–183 | Day 158 | Worksheet 79 | Name

On Cruise Control Along the Circulatory System Travel Brochure

You are the captain of a new cruise line called Circulating Cruise Capsules. Your mission is to take a group of tourists on the first successful navigation of the circulatory system.

Design a trifold cruise brochure to advertise this great adventure.

Format:

Cruise Line Name	Describe the cruise and the sights you will see on the way.	Schedule of travel.
Picture		
Catch Phrase	Cost:	
Date		

Drawing Comparisons

The upper part of the heart is the home of the atriums. The lower part of the heart is the home of the ventricles. Compare the right atrium to the left atrium and the right ventricle to the left ventricle. How are the chambers similar and different?

	Right Atrium	Left Atrium
Similar		
Different		

	Right Ventricle	Left Ventricle
Similar		
Different		

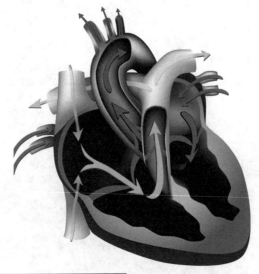

| The Complex Circulatory System | After Pages 184–185 | Day 159 | Activity 90 | Name |

What's for Lunch?

*REQUIRES ADULT SUPERVISION.

Materials:

- Pizza dough (You can purchase or make your own following any recipe you would like.)
- 2 tbsp. of olive oil
- Marinara or pizza sauce
- Sliced pepperoni
- Mozzarella or grated parmesan cheese
- Cookie sheet
- Non-stick cooking spray

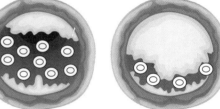

Early Atherosclerosis   Advanced Atherosclerosis   Severe Atherosclerosis

There is nothing better than enjoying your own finely crafted "artery." In this activity you will assemble your own "artery" pizza.

Instructions:

1. Ask an adult for help. Preheat oven to 400 degrees.

2. Spray your cookie sheet with non-stick cooking spray.

3. Take your dough and stretch it out in a flat cylinder shape. Then, lay it on your cookie sheet.

4. Brush the olive oil down the middle of the dough. Make sure to brush the oil the entire length of the pizza dough. This oil represents the fatty streak that forms in the arteries in the early stages of Atherosclerosis. Atherosclerosis is caused from excess fat, which deposits in the blood vessel. The choices you make in care for your body can have an effect on your artery health. Over time, things like poor eating habits, smoking, high blood pressure, and diabetes damage your vessels.

5. Spread marinara or pizza sauce on your doughy "artery." Make sure to be careful not to put the sauce all the way to the edge. Leave at least a half an inch from the edge of the dough unsauced.

6. Sprinkle the cheese on the lower half of your artery. This represents the plaques that can form in a damaged artery.

7. Add pepperoni to your model.

8. Place in oven. Cook until the pizza is golden brown, and the cheese is melted.

| The Complex Circulatory System | After Pages 184–185 | Day 160 | Worksheet 81 | Name |

The Heart

Use your book to help you label the following parts of the heart.

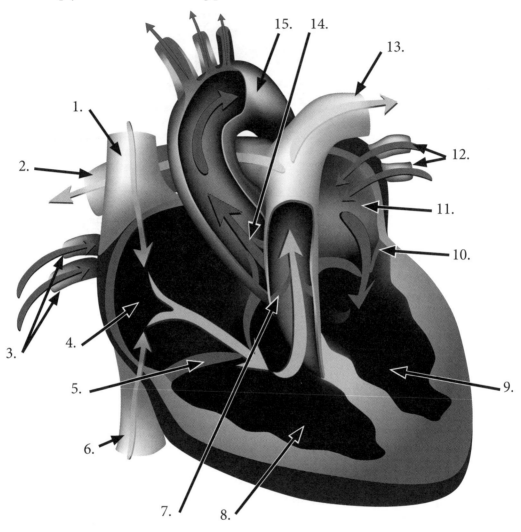

1. _____
2. _____
3. _____
4. _____
5. _____
6. _____
7. _____
8. _____
9. _____
10. _____
11. _____
12. _____
13. _____
14. _____
15. _____

314 // Elementary Anatomy, The Complex Circulatory System

| *The Complex Circulatory System* | After Pages 184–185 | Day 160 | Worksheet 82 | Name |

Getting the Blues Following the Leader

You will need red, blue, green, and brown coloring pencils or crayons to complete the following activity. Read the following instructions on the next page and follow what is requested.

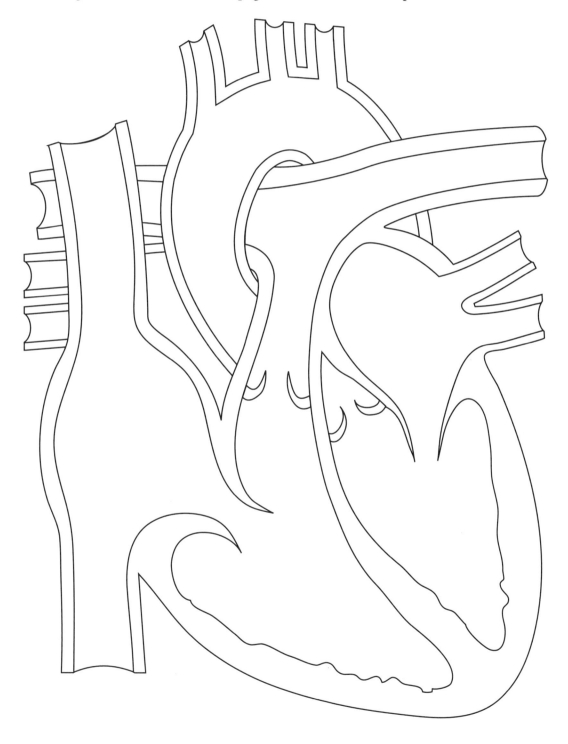

Instructions:

1. Label the valves of the heart. Color the flaps of the valves green.
2. Label the upper chambers of the heart.
3. Color the upper chamber in which deoxygenated blood is carried blue.
4. Color the upper chamber in which oxygenated blood is carried red.
5. Label the lower chambers of the heart.
6. Color the lower chamber in which deoxygenated blood is carried blue.
7. Draw arrows on the diagram that demonstrates the flow of blood.
8. The heart has its own electrical system. Draw a circle in the area in which the pacemaker of the heart is contained. Color it brown.
9. Oxygen-rich (oxygenated) blood flows through a vein from the lungs into the left atrium. Color this vein red.
10. Blood flows through an artery from the left ventricle and is pumped to the rest of the body. Color this artery red.
11. Oxygen-poor (deoxygenated) blood flows to the lungs from the right ventricle through an artery. Color this artery blue.

| The Complex Circulatory System | After Pages 186–187 | Day 161 | Activity 91 | Name |

The Highways of the Body

Materials:

- 5–6 ft. large butcher paper or white bulletin paper
- Markers
- Glue
- Scissors
- Red yarn
- Blue yarn
- White yarn
- Red construction paper
- Yellow construction paper

Instructions:

1. Lay the butcher paper on the floor. Lie on your back on top of the paper. Have someone trace the outline of your body onto the paper.

2. Cut out the pattern of the heart on page 319. Trace it onto the red construction paper.

3. Cut out the pattern of the lungs on pages 321 and 323. Trace it onto the yellow construction paper.

4. Glue the heart and lungs in the middle of the torso area. The heart should be in the center with a lung on each side.

5. Glue the red yarn throughout the body. The red yarn represents the arteries. Follow the pattern on page 325.

6. Glue the blue yarn throughout the body. The blue yarn represents the veins. Follow the pattern on page 325. This also represents the capillary networks throughout the body.

Page left intentionally blank.

Heart

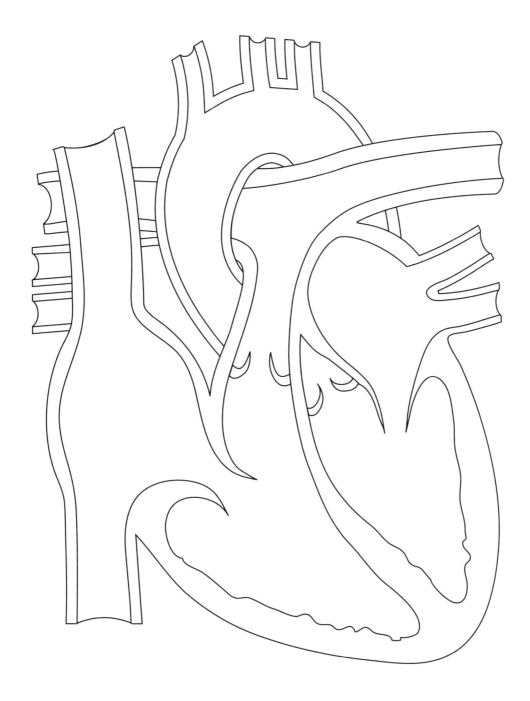

Activity 91, Day 161

Page left intentionally blank.

Right Lung

Page left intentionally blank.

Left Lung

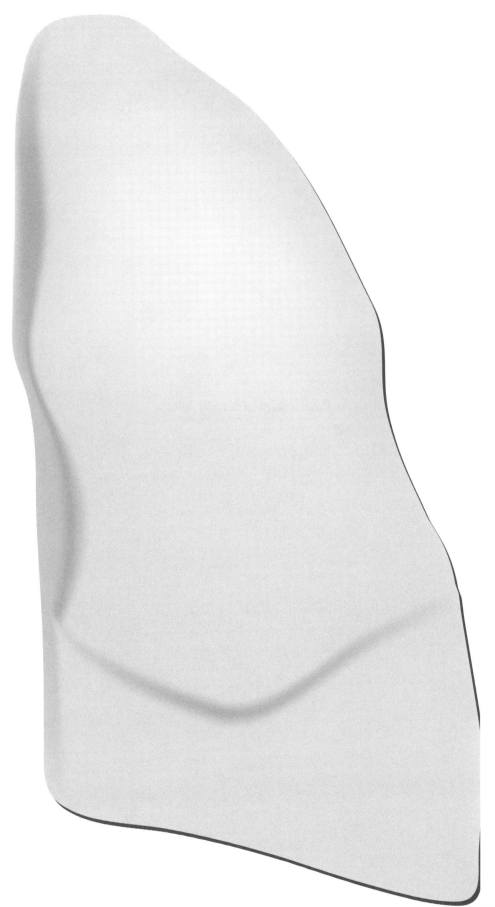

Page left intentionally blank.

Blood Vessel Diagram

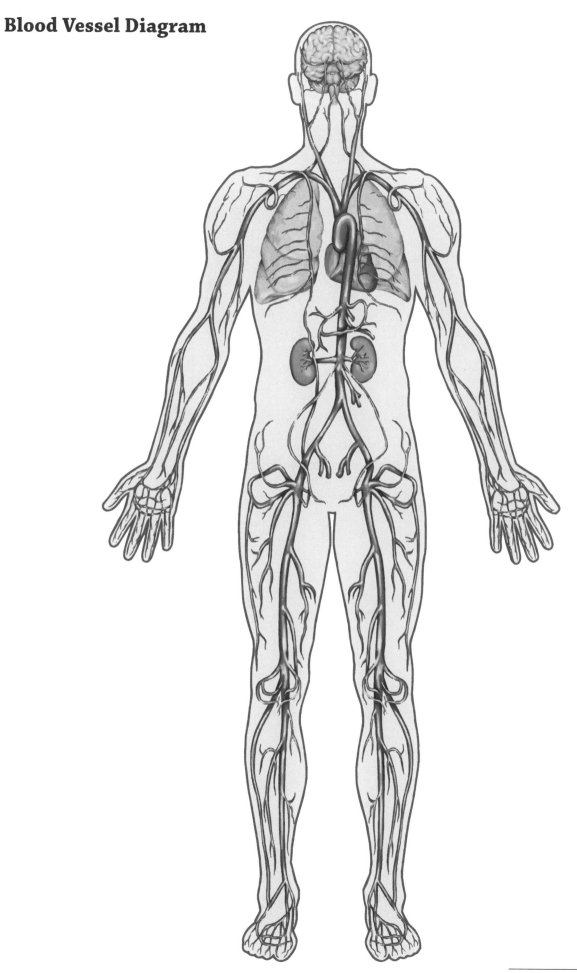

Page left intentionally blank.

| The Complex Circulatory System | After Pages 186–187 | Day 161 | Activity 92 | Name |

Feeling Under Pressure

As your heart beats, it forces the blood through the blood vessels at a certain pressure. The pressure is generated by the force of the heart. The stronger your heart is the more efficient it is at pushing your blood to the rest of your body. In this activity you will create a simulated heart. The object is to measure and record the distance your simulated heart pumps blood out of its chambers as different pressure is applied.

Materials:

- Squeeze bottle
- Meter or yard stick
- Water
- Red food coloring (optional)
  (Note: Be careful! This can stain skin and clothing.)

Instructions:

1. Go outside to a place on dirt or cement where you can squirt the liquid. You may add several drops of red food coloring and shake gently to simulate the blood.

2. Hold the squeeze bottle horizontal and use just one hand to squirt the first time. Record this measurement on the data chart. (Refill the bottle after each use.)

3. Next, squirt more firmly with two hands. Record this measurement on the data chart.

4. Finally, squirt with both hands firmly. Record this measurement on the data chart.

Data Chart

	Amount of Pressure Exerted	Distance
First squirt measurement		
Second squirt measurement		
Third squirt measurement		

1. At what point did you exert the most pressure?

2. How would you feel if you had to do this 100 times in a row? What if you had to do this 500 times in a row? What if you had to do this for 70 years?

3. Do you see how someone with high blood pressure would apply the greatest strain on their heart?

4. As you age, your blood pressure changes. A Sphygmomanometer, or blood pressure cuff, measures the pressure that blood surges through your arteries. There are two readings obtained and recorded from the cuff, the systolic and diastolic pressures.

   Look up the following terms. Write your answers below.

   Diastolic:

   Systolic:

What are the average blood pressures for a newborn and for an adult?

Newborn _____

Adult _____

| The Complex Circulatory System | After Pages 186–187 | Day 161 | Activity 93 | Name |

Are You As Strong As Your Heart?

Materials:

- Tennis ball
- Stopwatch or clock with a second hand
- Partner

Instructions:

1. Hold a tennis ball in your strongest hand.

2. Have your partner keep time for you.

3. At your partner's command begin to squeeze the ball eight times in five seconds.

4. Were you able to squeeze the ball that many times? This is how many times your heart squeezes in 5 seconds for a younger child.

The smaller you are, the faster your heart beats!

Age (years)	Heart Rate (beats/min)
<1	100–160
1–2	90–150
2–5	80–140
6–12	70–120
>12	60

Listen to My Heart

Adult supervision required.

Materials:

- 2 funnels
- A large balloon
- Rubber tubing
- Scissors
- Drinking straw
- Modeling clay

Instructions:

Part 1

1. Insert the tubing into the bottom of the funnels.

2. Wrap the tape around the tubing, affixing it to the funnels if needed.

3. Snip the open end of the balloon. Stretch the balloon over the opening of one of the funnels. Tape the balloon to keep it in place.

4. Your homemade stethoscope is ready to use.

5. What sounds do you hear?

_____

_____

Part 2

1. Find your pulse on your wrist with fingers below the prominent area under the thumb region of your hand.

2. Place a wad of modeling clay on top of the pulsing area.

3. Insert a drinking straw into the center of the modeling clay on your wrist. It should point straight up.

4. Rest your arm comfortably on the table. Listen to your heart while you watch the tip of the straw.

5. What are your observations?

_____

_____

Who Am I ?

Write the word next to the statement that describes it.

Word Bank
Blood   Heart   Blood Vessel

1. _____ I pump blood to all parts of the body.

2. _____ I am about the size of your fist.

3. _____ When you cut your skin, I form a sticky lump called a clot.

4. _____ I carry blood to different parts of the body.

5. _____ I am made up of tiny cells floating in a liquid.

6. _____ I beat more than 100,000 times a day.

7. _____ I form a system of tubes that would wrap around the earth two and half times.

8. _____ I carry oxygen to parts of the body.

9. _____ A strong outer covering protects me from rubbing against the lungs.

10. _____ A drop of me contains millions of red cells and thousands of white cells.

Take a Note

Your instructor will read page 188 in your textbook aloud. Listen carefully and fill in the blank spaces as the passage is being read aloud to you.

Your _____ began development as a simple _____. It began to _____, when you were in your _____ womb, at about _____ weeks.

The _____ circulation in a _____, a _____ in the womb, is different than the blood _____ in _____ and _____.

At _____, the baby's _____ _____ starts the miraculous conversion from _____ circulation to the baby _____ on its own.

The _____ that now fills the baby's _____ signals the _____ in the baby's body to _____ down.

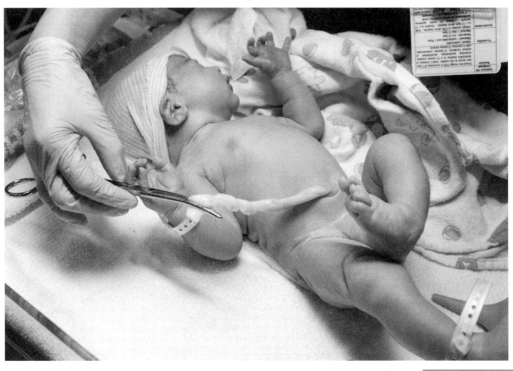

The da Vinci Robot

You are the Chief Executive Officer for a medical supply company that sells the da Vinci Surgical System. Your job is to create an advertisement for this amazing machine.

Your advertisement needs to include the following:

1. Title

2. Selling points within the advertisement

3. Color drawing of your invention/innovation

4. Description about why this invention/innovation is good for the hospital and patient

Title:

Price: Selling Points:

| The Complex Circulatory System | After Pages 192–193 | Day 163 | Worksheet 86 | Name |

Hands as Steady as a Surgeon

Cut out the parts of the body and paste them in their proper place on one (or both) of the children below.

Page left intentionally blank.

| | *The Complex Circulatory System* | After Pages 192–193 | Day 163 | Activity 95 | Name |

The Heart and Lungs Team Up

Your heart and lungs make a powerful team.

Put your hand on your chest.

Can you feel your heart beat?

Place both hands on your chest.

Can you feel your chest move when you breathe?

Your heart and lungs are connected.

The heart pushes blood into the lungs.

Your blood gets rid of carbon dioxide.

It collects oxygen in the lungs.

Materials:

- 2 sandwich bags
- 3 bag ties
- Small Styrofoam™ ball or balloon
- Red modeling clay

Instructions:

1. Take each of the sandwich bags and blow them up. Tie each bag off with a bag tie.
2. Use the third tie to connect the two air-filled bags together.
3. Blow up the balloon to the size of your fist. Tie the balloon off.
4. Take the Styrofoam™ or balloon (whichever you are using) and place it into the clay. Mold the clay over the object. This represents the heart.
5. Roll out two cylinder-shaped clay forms and place on top of the "heart." This represents the two major vessels that run in and out of the lungs.
6. Place the "lungs" on top with the major vessels. You now have a model of the powerful duo.

Page left intentionally blank.

Open Heart Surgery

REQUIRES ADULT SUPERVISION. Lay knife on the table when done. Remember, when working with sharp items, ALWAYS have adult supervision and keep away from younger students.

Materials:

- Sheep or cow heart
- Butcher knife
- Probe (optional)
- Meat tray or hard surface like a tray
- Gloves
- Sandwich bag

Instructions:

1. All good dissections begin with observation. Place the heart on your tray. Examine the blood vessels lining the outside of the heart. Note the fatty tissue located at the upper region of the heart. Why would the heart have fat in this region? Is fat always bad?

2. Identify the front (anterior) part of the heart from the back (posterior) of the heart. Clearly, from your observation the heart is not shaped in the classic valentine manner (♥). The heart is shaped asymmetrically. Place the vessels at the top. The front side of the heart will have a large vessel running in the coronary sinus diagonally through the center of the heart. The back side will be slightly flatter than the front side.

3. Position the heart with the front side up. Locate the auricle. The auricle is a flap of tissue, which can be lifted slightly up on the upper right part of the heart. Auricle means "ear." This part resembles an elephant ear, to a degree.

   The auricle acts as a run-off for extra blood that fills the right atrium and helps to control the pressure in the chamber. This is much like the expansion tank associated with many water heaters in homes.

4. Locate the four large vessels at the top of the heart. The vessels are typically clipped close to the preserved heart specimen. The four vessels to identify are the pulmonary artery and vein, the aorta, and superior vena cava.

5. Have an adult assist you with the next step. Lay the heart flat with the front side up. Place the palm of your non-dominant hand on top of the heart to keep it from sliding. Take a butcher knife, and beginning at the apex (the bottom of the heart) slice the heart into two sections. This slice will allow you to see all four chambers.

6. Locate the atriums and ventricles. Notice the difference in the thickness in the muscular walls between the ventricles. The left ventricles walls are much thicker than the right. Why do you think God designed it in this manner?

7. Locate the valves that separate the chambers of the heart. The valves are easily identified by the string or cords that attach to them. These are called the chordae tendineae and are

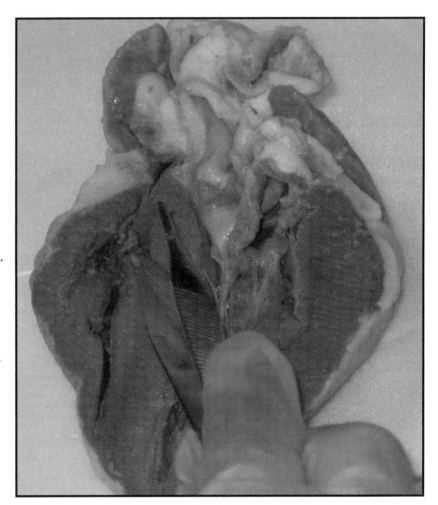

affectionately called the "heart strings." The chordae tendineae connects to the papillary muscles. These muscles resemble the tip of a finger. What is the function of the chordae tendineae?

8. Insert a pencil, straw, or other cylindrical object into the upper vessels of the heart. Identify where your object enters the heart. Can you determine which vessels are which?

9. Once you are done, you can dispose of your heart into the garbage or seal it in a sandwich bag for future examination. **Remember: do not place your specimen in water. It will destroy the specimen.** Keep your specimen at room temperature. There is no need to place the specimen in the freezer or refrigerator.

| The Complex Circulatory System | After Pages 194–195 | Day 164 | Worksheet 87 | Name |

Circulating Around and Crunching Numbers

There is an immense expanse of blood vessels and cells wrapped up in you. It is mind-boggling to even think about. Below are math calculations to perform based on *The Complex Circulatory System*.

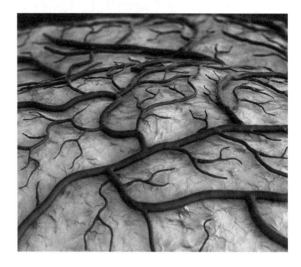

1. The average adult has within their person 4.5 L of blood. If 20 adults are waiting in the doctor's office, how many liters of blood would be circulating in total in all the persons combined?_____

   Show your work here:

2. Your blood volume is approximately 7 percent of your body weight. How much does the blood contained in your body weigh in kilograms?_____ Hint: 1 kg is approximately 2.2 pounds

   Show your work here:

3. When you are quietly resting, your heart will pump and circulate about 5 L of blood per minute. How many liters are pumped in six hours? _____ Two days? _____ In a week? _____

   Show your work here:

4. A newborn baby's heart beats on average 120 times per a minute. How many times does a baby's heart beat in 1 hour? _____ How many times in a month? _____

   Show your work here:

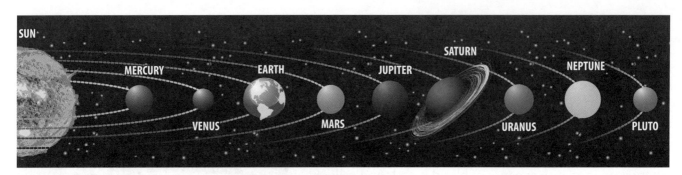

5. If you were to lay all the blood vessels in your body end to end it would stretch approximately 60,000 miles. If you were to tie all the blood vessels within several persons end to end, how many people would you need to reach Mars? _____ Saturn? _____ Neptune? _____

Distance from Earth to Mars = 128,000,000 miles (at its closest point to Earth)
Show your work here:

Distance from Earth to Saturn = 746,000,000 miles
Show your work here:

Distance from Earth to Neptune = 2,700,000,000 miles
Show your work here:

| The Complex Circulatory System | After Pages 194–195 | Day 165 | Activity 97 | Name |

Take a Stroll Through the Heart

Materials:

- Burlap drop cloth or old sheet (to protect the floor)
- Fabric markers
- Fabric paint: red, black, blue, pink
- Paint brush
- Plastic tablecloth (to protect the floor)
- Fabric pencil
- Yard stick
- Bean bag or sock stuffed with beans that is tied at the top

Instructions:

1. Lay the plastic tablecloth on the floor with the shiny side up.

2. Lay burlap or an old sheet on top of the tablecloth. This will ensure that the paint will not soak through the fabric and stain the floor.

3. Measure an area at the center of the fabric at least 3 ft. by 5 ft. This area can be bigger depending on the size of the fabric.

4. Mark the corners of the area off with a fabric pencil. Draw an evenly spaced grid that is 5x5 blocks using your yard stick and fabric pen. This grid method will assist you in drawing the picture of the heart in an equal ratio, just like the example on the next page. You simply focus on one square at a time and draw the line segments in the square, until you complete the image.

5. Once you have completed your line drawing, color the areas as indicated by number.

6. Once the coloring has been completed, outline all the sections in black.

7. Allow your masterpiece to dry overnight.

8. Your heart is now ready for a stroll. Follow the exploration questions as you walk through the heart. Enjoy.

Line Drawing of the Heart

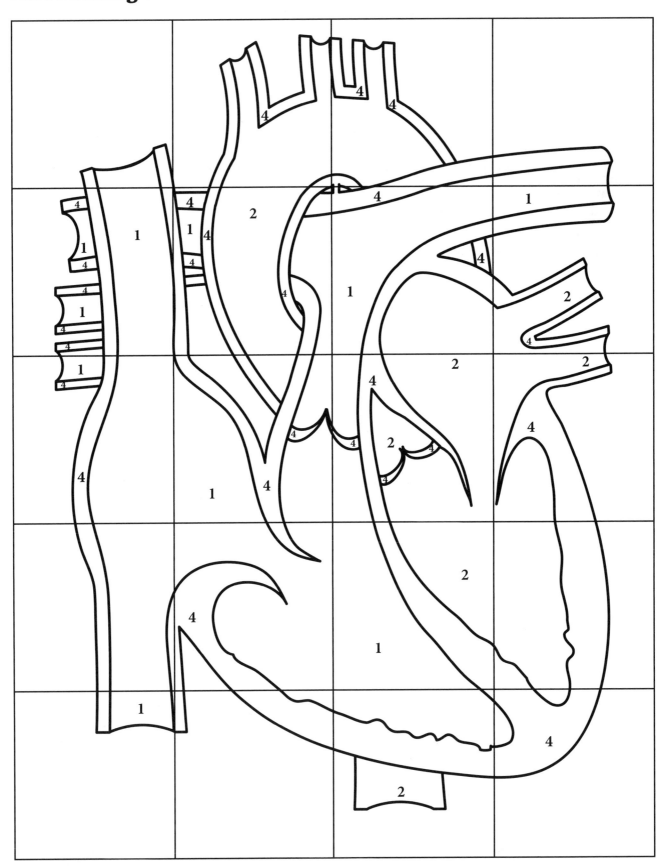

Paint by Number
1 - Blue   2 - Red   3 - Black   4 - Pink

Exploration Questions/Experiences:

Level 1

Play a game of Simon Says

Examples of Commands:

1. Jump three times in an Atrium.
2. Hope on one foot in a Ventricle.
3. Sit on the right side of the heart.
4. Lie down and roll to the left side of the heart.
5. Knock on a "doorway" between chambers of the heart.
6. Do four sit-ups in the area where the blood is rich in oxygen.
7. Do five jumping jacks in the area that has the least amount of oxygen in the blood.

Level 2

1. Stroll through the heart in the direction the flow of blood travels in the heart.
2. As you step on each space, name the structure you step on in the heart. (Students in this level would be expected to name at least the following structures: right atrium, right ventricle, pulmonary artery to the lungs, pulmonary vein, left atrium, left ventricle, and aorta)

Level 3

1. Stroll through the heart in the direction the flow of blood travels in the heart.
2. As you step on each space, name the structure you step on in the heart. (Students in this level would be expected to name all the following structures: superior vena cava, inferior vena cava, right atrium, tricuspid valve, right ventricle, pulmonic valve, pulmonary artery to the lungs, pulmonary veins, left atrium, mitral valve, left ventricle, aortic valve, and aorta)

Level 2 and 3

Determine and customize which questions to ask below based on the levels of your students. Have the student stand approximately 5 feet from the heart floor mural. Ask the student a question. They should take the bean bag/sock bag and toss it on the structure and verbally identify the correct answer to your questions.

Questions:

1. You may have noticed when you lay down on your back your abdomen pulses. What structure causes these pulsations? (Abdominal Aorta)
2. This is one of the structures that prevents the backflow of blood to the prior chamber. (valves)

3. This muscular wall divides the ventricles from each other. (septum)
4. Normally in the body, arteries carry oxygenated blood and veins carry deoxygenated blood. There are two exceptions to this rule in the heart. Toss your bean bag on one of these structures. (Pulmonary Artery or Pulmonary Vein)
5. This chamber must generate enough pressure to propel the blood to the entire body (left ventricle).
6. This structure receives blood from the lower part of the body. (Inferior Vena Cava)
7. This area contains blood that has the least amount of oxygen content. (Right Atrium and Ventricle)
8. What is the largest artery in the body? (Aorta)
9. In which chamber is the pacemaker of the heart located? (Right atrium)
10. This structure receives blood from the upper part of the body. (Superior Vena Cava)

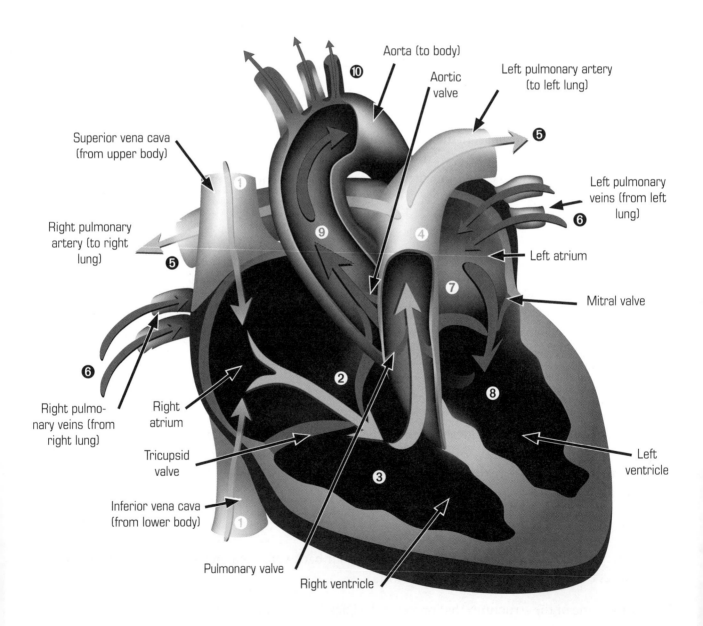

| The Complex Circulatory System | After Pages 196–197 | Day 166 | Worksheet 88 | Name |

Heart Health

God calls us to be good stewards. In 1 Corinthians 6:19–20 it states:

> *Do you not know that your bodies are temples of the Holy Spirit, who is in you, whom you have received from God? You are not your own; you were bought at a price. Therefore honor God with your bodies.*

God calls us to take care of his many gifts. Our bodies are one of these gifts. If we are not being good stewards of our bodies by putting junk into it, it will not work as effectively. If we do not feel well, it is hard to do the work God has called us to do.

Below you will see pictures of healthy and unhealthy things to put in or do with your body. Place a check next to things that are good for you. Cross out the items that over time can be very damaging to your body.

| The Complex Circulatory System | After Pages 196–197 | Day 166 | Activity 98 | Name |

You Are What You Eat: A Heart Healthy Diet

Materials:

- Old magazines
- Grocery store ads
- Assortment of foods
- Nutrition food labels

God has created a wonderful assortment of foods for our bodies to consume. When we put good foods into our bodies, our bodies run more efficiently. There are foods that we should not eat too often because over time these foods not only damage our bodies but our heart.

Instructions:

*Depending on the level of your student this activity can be done three different ways.

First Way:

1. Collect old magazines and/or grocery store food ads. Have your student cut out a variety of different foods.
2. Sort the foods into three groups. One group will represent foods that are good for your heart. The second group will represent foods that are bad for your heart if eaten too often. The third group will represent foods that you are not certain whether they are good or bad for your heart.
3. Once sorted, complete the follow-up questions on page 350.

Second Way:

Complete the above steps, but this time use actual foods around the house. Sort them into the above three groups. Once sorted, complete the follow-up questions on page 350.

Third Way:

1. Collect a wide assortment of food labels. Based on the information on the nutritional label sort the foods into the aforementioned three groups.
2. See the following page for how to read nutritional labels.
3. Once sorted, complete the follow-up questions on page 350.

Nutrition Facts

Serving Size 1 cup (228g)
Servings Per Container about 2

Amount Per Serving

Calories 250 Calories from Fat 110

	% Daily Value*
Total Fat 12g	18%
Saturated Fat 3g	15%
Trans Fat 3g	
Cholesterol 30mg	10%
Sodium 470mg	20%
Total Carbohydrate 31g	10%
Dietary Fiber 0g	0%
Sugars 5g	
Proteins 5g	
Vitamin A	4%
Vitamin C	2%
Calcium	20%
Iron	4%

* Percent Daily Values are based on a 2,000 calorie diet. Your Daily Values may be higher or lower depending on your calorie needs:

	Calories:	2,000	2,500
Total Fat	Less than	65g	80g
Saturated Fat		25g Less than	20g
Cholesterol	Less than	300mg	300mg
Sodium	Less than	2,400mg	2,400mg
Total Carbohydrate		300g	375g

For educational purposes only. This label does not meet the labeling requirements described in 21 CFR 101.9.

① Serving Size

This section is the basis for determining number of calories, amount of each nutrient, and %DVs of a food. Use it to compare a serving size to how much you actually eat. Serving sizes are given in familiar units, such as cups or pieces, followed by the metric amount, e.g., number of grams.

② Amount of Calories

If you want to manage your weight (lose, gain, or maintain), this section is especially helpful. The amount of calories is listed on the left side. The right side shows how many calories in one serving come from fat. In this example, there are 250 calories, 110 of which come from fat. The key is to balance how many calories you eat with how many calories your body uses. *Tip: Remember that a product that's fat-free isn't necessarily calorie-free.*

③ Limit these Nutrients

Eating too much total fat (including saturated fat and trans fat), cholesterol, or sodium may increase your risk of certain chronic diseases, such as heart disease, some cancers, or high blood pressure. The goal is to stay below 100%DV for each of these nutrients per day.

④ Get Enough of these Nutrients

Americans often don't get enough dietary fiber, vitamin A, vitamin C, calcium, and iron in their diets. Eating enough of these nutrients may improve your health and help reduce the risk of some diseases and conditions.

⑤ Percent (%) Daily Value

This section tells you whether the nutrients (total fat, sodium, dietary fiber, etc.) in one serving of food contribute a little or a lot to your total daily diet.

The %DVs are based on a 2,000-calorie diet. Each listed nutrient is based on 100% of the recommended amounts for that nutrient. For example, 18% for total fat means that one serving furnishes 18% of the total amount of fat that you could eat in a day and stay within public health recommendations. Use the Quick Guide to Percent DV (%DV): 5%DV or less is low and 20%DV or more is high.

⑥ Footnote with Daily Values (DVs)

The footnote provides information about the DVs for important nutrients, including fats, sodium, and fiber. The DVs are listed for people who eat 2,000 or 2,500 calories each day.

The amounts for total fat, saturated fat, cholesterol, and sodium are maximum amounts. That means you should try to stay below the amounts listed.

Follow-up Questions:

1. Discuss how you made the determination of what foods belonged in which group. Do you find that your diet contains a good deal of these types of foods?

2. Give your own definition of what it means to have a "heart-healthy" diet.

3. Did you place foods in the third group? What made the decision for each of the foods you selected to be placed here? After talking about the foods with your instructor or classmates, where will you now place the foods?

Scrambled Sentences

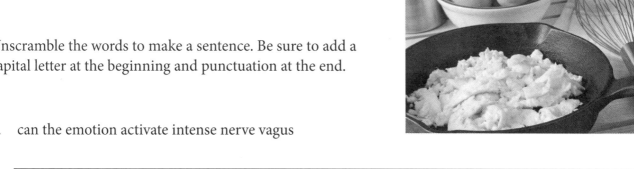

Unscramble the words to make a sentence. Be sure to add a capital letter at the beginning and punctuation at the end.

1. can the emotion activate intense nerve vagus

2. blood the backwards veins valves flowing keep help to from in

3. blood help chest blood compressions keep to circulating the

4. of is hardening arteriosclerosis the called arteries

Bruise in Sequence

Number the pictures in order of the development and stages of a bruise.

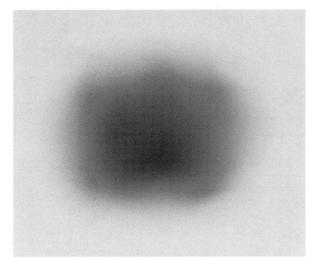

Bruise turns green to yellow, WBCs clean up the mess and dispose of decaying RBCs

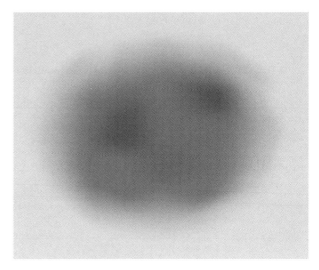

Red, swelling

Light brown, gets lighter and lighter

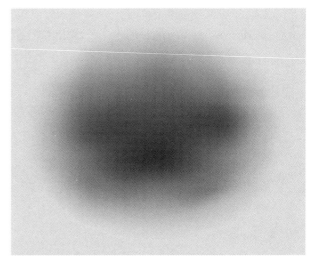

Swelling decreases; blue, black, and deep purple; RBCs begin to decay

Solve the Fact

Can you find out what the interesting fact about the heart resides below? Find the sum of each of the number pairs. Write the code letter of each answer on the correct line.

3 + 4 = _____ = A   5 + 7 = _____ = B   1 + 3 = _____ = C

0 + 9 = _____ = D   1 + 5 = _____ = E   3 + 10 = _____ = F

0 + 0 = _____ = G   5 + 6 = _____ = H   0 + 2 = _____ = I

5 + 5 = _____ = J   2 + 1 = _____ = K   1 + 0 = _____ = L

9 + 9 = _____ = M   6 + 8 = _____ = N   9 + 8 = _____ = O

1 + 4 = _____ = P   11 + 12 = _____ = Q   5 + 3 = _____ = R

9 + 7 = _____ = S   10 + 10 = _____ = T   12 + 9 = _____ = U

8 + 7 = _____ = V   10 + 9 = _____ = W   12 + 12 = _____ = Y

___ ___ ___ ___ ___ ___ ___ ___   ___ ___ ___ ___ ___ ___
 7   3   2 20   4 11   6 14   13   7 21   4   6 20

___ ___ ___ ___ ___   ___ ___ ___ ___   ___ ___
 19 17 21   1   9   14   6   6   9   20 17

___ ___   ___ ___ ___ ___   ___ ___   ___ ___   ___ ___ ___
 12   6   20 21   8 14   6   9   17 14   13 17   8

at least 45 years to equal the amount of blood pumped by the heart in an average lifetime.

| *The Complex Circulatory System* | After Pages 202–203 | Day 169 | Worksheet 92 | Name |

You've Got a Lotta Heart

Solve the following clues. Use your clues to complete the puzzle on the next page.

Find your clues in the word search on the next page. Words can be found backwards, diagonally, vertically and horizontally.

1. The phase in the heart cycle of beating in which the chambers contract and push blood out of the heart: _____.

2. The _____ coat is the middle section of a centrifuged sample of blood.

3. _____ cardia is a fast heartbeat, which typically is over 100 beats a minute.

4. The Pericardial _____ is fibrous and double-layered surrounding the heart.

5. _____ globin is an iron within the red blood cell that helps carry oxygen.

6. To lose consciousness or pass out: _____.

7. To listen to the sounds of the body: _____.

8. _____ is a problem with blood in which oxygen delivered to the organs is decreased.

9. The blood vessels in the body that carry deoxygenated blood to the heart are called _____.

10. A disease transmitted by mosquitos and to which people with Sickle Cell Anemia are immune: _____.

11. The blood vessels that line the outside of the heart are called _____ arteries.

There are other cardiovascular words lurking in this puzzle. How many additional ones can you find? Circle them in the puzzle and write them below:

```
A T R O A C A R D I U M
R C A P I L L A R Y B A
T E O D I A S T O L E L
E N D R S Y N C O P E A
R T B G O F L C K P Q R
I R G D O N O R A I M I
E I C E J V A L V E S A
S F E A S E I R E T R A
S U A N E M I A Y I T S
E G D T V N S A J R A E
T E R I A L T H I E C I
A T O G L M E U D J H D
T U H E H E M O K L Y O
L N C N F L N Y F F U B
U M E S N I E V E W K I
C E S B O R H F T L D T
S N H A G A J B U I P N
U O E L O T S Y S A C A
A B R A D Y C A R D I A
```

| The Complex Circulatory System | After Pages 202–203 | Day 170 | Worksheet 93 | Name |

Got the Smarts Real Good

List three new things that you learned during your reading of *The Complex Circulatory System*.

1. _____

2. _____

3. _____

Draw a picture or diagram of one of these things you learned.

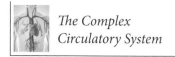

The Complex Circulatory System | After Pages 204–207 | Day 171 | Worksheet 94 | Name

Breaking the Ice with Idioms

An idiom is a phrase or expression that has a meaning different from what the words suggest in their usual meaning. Pick out two of your favorite idioms about the heart or blood and write the expression below. Draw a picture of the actual meaning of the idiom in the left picture box. In the box next to your idiom draw a literal picture of the expression.

Idiom Expression:_____

Meaning of Idiom	Literal Picture

Idiom Expression:_____

Meaning of Idiom	Literal Picture

Page left intentionally blank.

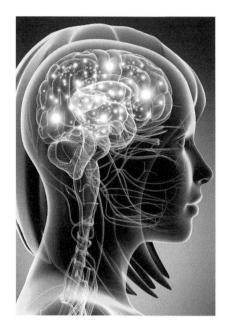
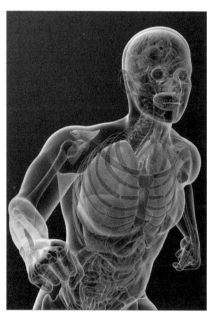
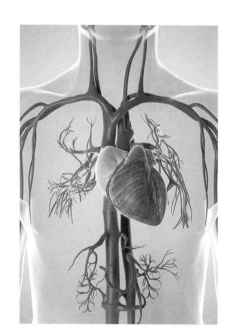

Portfolio / Rubrics / Reports
for Use with
*Elementary Anatomy: Nervous,
Respiratory, and Circulatory Systems*

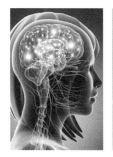
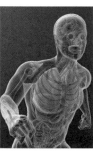
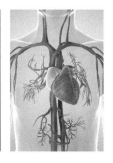

PORTFOLIO/RUBRICS /REPORTS

A. Biography Rubric

Biography Writing Rubric

	Score			
	1	**2**	**3**	**4**
Writing Ideas: Includes interesting and informative details	No details demonstrated in the writing	Writing has only a few details that support the main idea	Writing demonstrates many details that support the main idea	Quality details demonstrated in interesting and informative ways
Organization of Report: Report demonstrates a beginning and ending	No beginning or ending present at all in the writing	Vague beginning and ending demonstrated	Writing demonstrated a beginning, a middle, and an end with transitional sentences	Writing had a clear beginning, middle, and end; excellent transitional sentences and thought flowed smoothly
Word Choice: Use of descriptive and illustrative language	No adjectives or descriptive words used in writing	Writing had a few descriptive words; writing lacked variety or style	Many descriptive words; word choice at times took away from the meaning	Colorful and precise descriptive words that really paint a picture in the reader's mind
Spelling and Grammar	More than 10 spelling, punctuation, and grammar errors	Up to 10 spelling, punctuation, and grammar errors	Less than five spelling, punctuation, and grammar errors	Only one or two spelling, punctuation, and grammar errors

Comments:

Score: (Tally the score from each criterion)

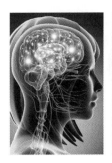
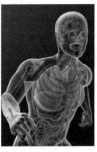
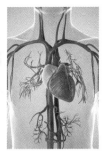

PORTFOLIO/RUBRICS/REPORTS

B. Oral Rubric

Oral Report Rubric

	Score 1	Score 2	Score 3	Score 4
Content of Oral Presentation: Relates to topic, detailed and precise	Most of the information was the writer's opinion; very few facts	Demonstrated basic understanding of the topic; many opinions were not supported by facts	Content was supportive of the topic, and almost all the opinions were supported by facts	Excellent supportive facts of the opinions expressed; each statement supported by a fact
Knowledge of the topic	Demonstrated little or no knowledge of the topic; had no supporting facts or details	Demonstrated basic knowledge of the topic; had one or two facts	Demonstrated a working knowledge of the topic; had three or more details	Excellent display of knowledge of the topic with many facts and details
Stays on Topic about the subject of presentation	More than half the presentation did not relate to the subject	At least half of the presentation did relate to the subject	Most of the presentation related to the subject	The entire presentation related to the subject
Audience Response: Engaging and interactive with audience	Audience lost interest; presenter was unable to answer questions	Audience interested at least half the time of the presentation; presenter tried to answer audience questions	Audience engaged for most of the presentation; presenter answered most of the questions	Audience engaged the entire presentation; presenter answered all questions
Posture and Eye Contact	Poor posture and made no eye contact	Posture was not good and made minimal eye contact	Posture was good and eye contact was made for much of the presentation	Excellent posture and eye contact throughout the presentation

Comments:

Score: (Tally the score from each criterion)

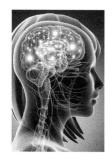
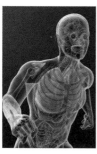
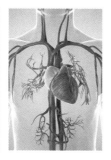

PORTFOLIO/RUBRICS/REPORTS

C. Science Experiment Rubric

Science Experiment Rubric

	Score 1	Score 2	Score 3	Score 4
Hypothesis = Prediction between experiment and results	No hypothesis	No connection between the hypothesis and experiment	Hypothesis and the problem were clearly connected	Excellent connection between the problem and expected outcome of experiment was obvious
Lab Work Observations: Record data about what happened	Missed more than three parts of the data; wrote some of the steps of the experiment	Missed up to three parts of data; wrote down most of the steps in the experiment	Wrote down all data; had details about what transpired during the experiment	Wrote down all data clearly, with many details and illustrations
Answers and predictions relate to the hypothesis	Did not answer prediction or did not relate to the hypothesis	Did not draw a conclusion about the hypothesis from the experiment	Answered predictions and wrote whether hypothesis was true or false based on experiment	Was able to draw conclusions from the experiment; if the hypothesis was false, the student changed and a new experiment was proposed

Comments:

Score: (Tally the score from each criterion)

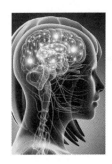

PORTFOLIO/RUBRICS/REPORTS

D. Objective Concept Checklist
Objective Concept for the Nervous System Checklist

Rating Scale:

1. No knowledge of concept
2. Limited understanding
3. Good understanding of concept
4. Excellent understanding of concept

Concept	1	2	3	4
Name the major regions of the brain and describe their functions				
Identify the location and composition of gray and white matter				
Locate the cerebral hemispheres				
Name the three divisions of the diencephalon				
Explain how the brain is located, supported, and protected in the cranial vault				
Explain the blood-brain barrier				
Locate the sensory, motor, and association areas of the cerebral cortex and discuss their functions				
Identify the important structures within the regions of the brain and explain their prospective functions				
Identify the gross anatomical features of the spinal cord				
Understand and explain dermatomes				

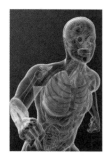

PORTFOLIO/RUBRICS/REPORTS

E. Objective Concept Checklist

Objective Concept for the Respiratory System Checklist

Rating Scale:

1. No knowledge of concept
2. Limited understanding
3. Good understanding of concept
4. Excellent understanding of concept

Concept	1	2	3	4
Describe the primary function of the respiratory system				
Describe the function of mucus				
Name the location of the sinuses and their function				
Name the respiratory structures and describe their functions				
Describe the function of the nasal cavity and nostril hairs				
Name the parts of the upper respiratory tract				
Name the parts of the lower respiratory tract				
Name the tissue comprising the rigid support of the larynx				
Describe the function of the epiglottis and vocal cords				
Describe the location and function of bronchi, bronchioles, and alveoli				
Describe the function of cilia				
Describe the function of the diaphragm				
Describe how gas exchange occurs between the alveoli and pulmonary capillaries				
Describe where the main respiratory control center is located				
Describe how oxygen is transported throughout the body				

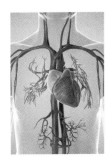

PORTFOLIO/RUBRICS/REPORTS

F. Objective Concept Checklist
Objective Concept for the Circulatory System Checklist

Rating Scale:

1. No knowledge of concept
2. Limited understanding
3. Good understanding of concept
4. Excellent understanding of concept

Concept	1	2	3	4
Identify at least three themes from the Bible demonstrating the significance of our "hearts."				
Identify the functions of blood.				
Name the different parts of the blood.				
Explain the function of each blood cell.				
Explain the basic process of a blood clot formation.				
Identify the basic blood types.				
Explain the donor/recipient relationship between the various blood types.				
Compare and contrast the similarities and differences between arteries and veins.				
Name the five types of vessels in the body as blood flows through the complete circuit from the left side of the heart around the body to the right side of the heart.				
Explain the difference between open and closed circulatory systems.				
Locate and label structures of the heart in a diagram.				
Trace the pathway of blood in the heart using arrows on the diagram.				
Recognize the attributes of the heart structure as a double pump system (right and left side).				
Define the terms systole and diastole.				
Discuss heart sounds in terms of what they represent and how they sound.				

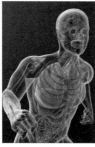
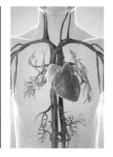

PORTFOLIO/RUBRICS/REPORTS

G. Lab Reports

Scientific Lab Report

Title:_____

Date:_____

Ask a Question or Discuss the Problem . . .

State Your Hypothesis . . .

Materials List

1.

2.

3.

4.

Procedure:

Observations:

Conclusions:

Observational Drawings:

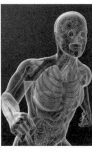
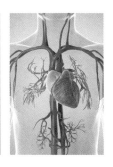

PORTFOLIO/RUBRICS/REPORTS

H. Learning Log

My Learning Log

Activity: Date:

My goals for today were:

I accomplished the following things today:

I still have questions about:

The next steps I need to take:

I learned:

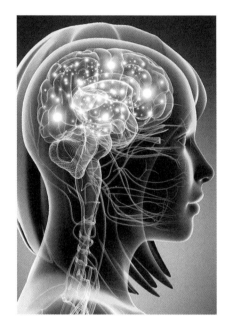
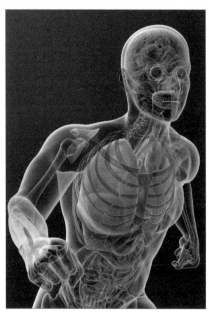
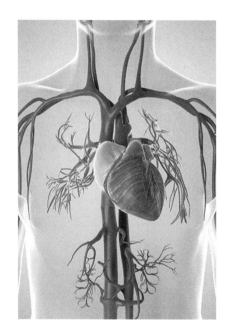

Appendices
for Use with
*Elementary Anatomy: Nervous,
Respiratory, and Circulatory Systems*

The Electrifying Nervous System

ACTIVITY/WORKSHEET OVERVIEW

Activity (ACT) / Worksheets (WS)	Description	Intrapersonal	Verbal–Linguistic	Logical–Math	Visual–Spacial	Body–Kinesthetic	Musical–Rhythmic	Interpersonal	Naturalist	1	2	3
ACT1	Nervous System Flash Cards		✓			✓				✓	✓	✓
ACT2	Back in Time	✓									✓	✓
ACT3	How Did it Happen?	✓									✓	✓
ACT4	Supercilious							✓				✓
ACT5	Techy		✓								✓	✓
ACT6	Timeline Shuffle				✓	✓					✓	✓
ACT7	Neuron Connections				✓					✓	✓	✓
ACT8	You've Gotta Nerve			✓	✓						✓	✓
ACT9	Jingle, Jangle						✓				✓	✓
ACT10	Brain Transplant	✓										✓
ACT11	Head Injuries	✓								✓	✓	✓
ACT12	Blood-Brain Barrier				✓							✓
ACT13	Write It Out			✓	✓						✓	✓
ACT14	Act It Out			✓	✓	✓					✓	✓
ACT15	Little Man in the Brain				✓	✓					✓	✓
ACT16	Broca / Wernicke							✓				✓
ACT17	Penelope's Lost Puppy	✓	✓									
ACT18	Drug Effects		✓									✓
ACT19	Brainiac			✓	✓					✓	✓	✓
ACT20	Training Your Cerebellum					✓				✓	✓	✓
ACT21	Brain Disorders				✓							✓
ACT22	Dough Brain Model				✓	✓				✓	✓	✓
ACT23	Egghead				✓	✓				✓	✓	✓
ACT24	Holes in the Head		✓									✓
ACT25	A Bird on a Wire		✓					✓				✓
ACT26	Stem Cells		✓	✓								✓
ACT27	Spinal Cord					✓		✓				✓
ACT28	Spinal Column			✓							✓	✓
ACT29	If You Couldn't				✓	✓						
ACT30	Reflexes					✓				✓	✓	✓
ACT31	It is All in the Timing					✓					✓	✓
ACT32	Light as a Feather					✓				✓	✓	✓
ACT33	The Scary Stuff Challenge	✓	✓		✓							

Activity (ACT) / Worksheets (WS)	Description	Multiple Intelligence Type									Ability Level		
		Intrapersonal	Verbal–Linguistic	Logical–Math	Visual–Spacial	Body–Kinesthetic	Musical–Rhythmic	Interpersonal	Naturalist	1	2	3
ACT34	Memory				✓					✓	✓	✓
ACT35	Take a Picture				✓					✓	✓	✓
ACT36	Fun with Limericks						✓				✓	✓
ACT37	Sleep	✓								✓	✓	
ACT38	Sleep Simulator	✓										
ACT39	Lame Brains			✓						✓	✓	✓
ACT40	Brain Strain			✓						✓	✓	✓
ACT41	Shrunken Apple Head				✓	✓					✓	✓
ACT42	I am Wonderfully Made Part 1				✓	✓						
ACT43	I am Wonderfully Made Part 2				✓	✓						
ACT44	Brain Salad			✓	✓						✓	✓
ACT45	Brain Lab					✓			✓		✓	✓
WS1	Just the Facts	✓								✓	✓	✓
WS2	Biblical References #1	✓					✓					✓
WS3	Biblical References #2		✓							✓	✓	✓
WS4	The Word of God		✓								✓	✓
WS5	Looking Inside the Brain		✓									
WS6	Back to Basics		✓									✓
WS7	Basics of the Nervous System				✓					✓	✓	✓
WS8	Word Scramble: Major Regions	✓								✓	✓	✓
WS9	Anatomy of the Central Nervous System				✓					✓	✓	✓
WS10	Cerebrum		✓								✓	✓
WS11	Type of Brain						✓					✓
WS12	Regions of the Brain				✓					✓	✓	✓
WS13	Action and Control		✓								✓	✓
WS14	Frontal Lobe		✓								✓	✓
WS15	Life of Dr. Penfield		✓									✓
WS16	Homunculus				✓						✓	✓
WS17	Brain and Growth	✓	✓	✓								✓
WS18	Fearfully and Wonderfully Made		✓	✓							✓	✓
WS19	Blood-Brain Barrier Maze	✓								✓	✓	✓
WS20	The Backbone		✓							✓	✓	✓
WS21	All Circuits Firing		✓									
WS22	The Hypothalamus					✓				✓	✓	✓
WS23	A Personal Sleep Study					✓				✓	✓	✓
WS24	Brain Food					✓				✓	✓	✓
WS25	Cross Over		✓									✓
WS26	Do You Know?	✓						✓		✓	✓	✓

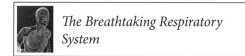
The Breathtaking Respiratory System

ACTIVITY/WORKSHEET OVERVIEW

Activity (ACT) / Worksheets (WS)	Description	Intrapersonal	Verbal–Linguistic	Logical–Math	Visual–Spacial	Body–Kinesthetic	Musical–Rhythmic	Interpersonal	Naturalist	1	2	3
ACT46	Respiratory Flash Cards		✓			✓				✓	✓	✓
ACT47	Snot O' Matic				✓	✓				✓	✓	✓
ACT48	Wind Bag			✓		✓					✓	✓
ACT49	Adventures of Jack Haldane				✓							
ACT50	Measuring Lung Capacity				✓	✓				✓	✓	✓
ACT51	Smoke in Your Lungs				✓	✓			✓			✓
ACT52	What's the Count?			✓				✓				✓
ACT53	The Voice of a Balloon						✓			✓	✓	
ACT54	Breathe					✓				✓		
ACT55	Wind Bag					✓				✓	✓	
ACT56	What Is Your Girth?						✓			✓	✓	
ACT57	Uniqueness of the Lungs		✓		✓							
ACT58	A Nose for Clues					✓			✓	✓	✓	✓
ACT59	Grape-Like Clusters				✓					✓	✓	✓
ACT60	Up in Smoke							✓				✓
ACT61	Pulse and Respiratory Rates					✓		✓			✓	✓
ACT62	The Air Around Us							✓		✓	✓	✓
ACT63	No Idling	✓						✓				✓
ACT64	What Trees Want				✓			✓			✓	✓
ACT65	Spanish Flu Pandemic of 1918	✓										
ACT66	A Very Simple Lung Model				✓					✓	✓	✓
ACT67	Know Your Noses			✓		✓				✓		
ACT68	Sticky Situations	✓					✓					✓
ACT69	Picture This	✓			✓	✓					✓	✓
ACT70	The Doctor Is In	✓	✓									✓
ACT71	Incentive Spirometer					✓					✓	✓

Activity (ACT) / Worksheets (WS)	Description	Multiple Intelligence Type								Ability Level		
		Intrapersonal	Verbal–Linguistic	Logical–Math	Visual–Spacial	Body–Kinesthetic	Musical–Rhythmic	Interpersonal	Naturalist	1	2	3
ACT72	Sing Your Heart Out						✓			✓	✓	✓
ACT73	It's a Record!		✓									
ACT74	It's Just a Game	✓		✓							✓	✓
WS27	The Word of God		✓					✓				✓
WS28	Timeline Shuffle				✓	✓					✓	✓
WS29	Biblical Reference		✓							✓	✓	
WS30	Dissenter / Sufferers		✓								✓	✓
WS31	Basics of the Respiratory System			✓	✓					✓	✓	✓
WS32	ID Parts of the Nasal Cavity		✓	✓							✓	✓
WS33	ID Parts of the Respiratory System		✓	✓							✓	✓
WS34	How We Breathe				✓							
WS35	Build a Word				✓							
WS36	Don't Be Cross		✓									✓
WS37	Life of an Oxygen Molecule		✓									✓
WS38	What's Your Function?		✓	✓						✓	✓	
WS39	Inhale, Exhale		✓	✓						✓	✓	
WS40	Digging Deeper	✓	✓									✓
WS41	Scrambled Eggs Level 1		✓							✓		
WS42	Scrambled Eggs Level 2		✓								✓	
WS43	Scrambled Eggs Level 3		✓									✓
WS44	Every Breath You Take				✓							✓
WS45	Lung Damage				✓							
WS46	Good Habits for Health	✓										
WS47	Fill in the Blanks	✓										
WS48	Puzzle Fun				✓							
WS49	Respiratory Problems					✓						
WS50	It's a Secret!				✓							
WS51	Martha and Wilma		✓									
WS52	Ponder This!	✓										
WS53	Do You Know?	✓								✓	✓	✓

The Complex Circulatory System

ACTIVITY/WORKSHEET OVERVIEW

Activity (ACT) / Worksheets (WS)	Description	Intrapersonal	Verbal–Linguistic	Logical–Math	Visual–Spacial	Body–Kinesthetic	Musical–Rhythmic	Interpersonal	Naturalist	1	2	3
		Multiple Intelligence Type								Ability Level		
ACT75	Flash Cards		✓		✓					✓	✓	✓
ACT76	Cardiac Cobble Stones		✓		✓	✓				✓		
ACT77	Historical Timeline			✓	✓						✓	✓
ACT78	Blood Splatter Art				✓	✓				✓	✓	✓
ACT79	The Extraordinary Journey	✓	✓		✓						✓	✓
ACT80	How Much Blood?	✓			✓	✓				✓		
ACT81	Simple Fake Blood	✓				✓				✓	✓	✓
ACT82	Easy Blood Model	✓		✓	✓					✓	✓	✓
ACT83	Another Easy Blood Model	✓		✓	✓					✓	✓	✓
ACT84	Multi-Sensory Hands–on Blood	✓			✓	✓				✓	✓	
ACT85	Blood Transfusion Simulation	✓		✓	✓	✓					✓	✓
ACT86	Test Your Knowledge on Blood											
ACT87	Match a Blood Sucker			✓	✓					✓	✓	
ACT88	A Slimy Situation	✓		✓	✓						✓	✓
ACT89	A Road Block	✓		✓	✓	✓					✓	
ACT90	What's for Lunch?	✓		✓		✓					✓	✓
ACT91	The Highways of the Body	✓			✓	✓				✓	✓	
ACT92	Feeling Under Pressure			✓	✓	✓					✓	✓
ACT93	Are You as Strong as Your Heart?	✓				✓				✓	✓	
ACT94	Listen to My Heart	✓				✓	✓			✓	✓	✓
ACT95	The Heart and Lungs Team Up				✓	✓				✓	✓	
ACT96	Open Heart Surgery	✓		✓	✓			✓			✓	✓
ACT97	Take a Stroll Through the Heart				✓	✓			✓	✓	✓	✓
ACT98	You Are What You Eat: A Heart Healthy Diet	✓			✓	✓				✓	✓	✓

Activity (ACT) / Worksheets (WS)	Description	Multiple Intelligence Type									Ability Level		
		Intrapersonal	Verbal–Linguistic	Logical–Math	Visual–Spacial	Body–Kinesthetic	Musical–Rhythmic	Interpersonal	Naturalist	1	2	3
WS54	Copywork: Cursive Prov. 15:13–14		✓								✓	
WS55	Copywork: Manuscript Prov. 15:13–14		✓							✓		
WS56	Heart Matters and Its Meaning		✓	✓								✓
WS57	Circulatory System Report		✓					✓				✓
WS58	Early Ideas on the Circulatory System			✓	✓						✓	✓
WS59	Help Wanted Ad		✓	✓				✓				✓
WS60	Parts of the Blood		✓		✓						✓	✓
WS61	Put Some Color in It			✓	✓					✓		
WS62	Blood Smear				✓			✓		✓	✓	
WS63	What Color Is Your Blood?											
WS64	Put a Plug in It			✓	✓						✓	✓
WS65	Color Your Heart		✓		✓					✓		
WS66	The Blood Report		✓		✓			✓				✓
WS67	You're Not My Type				✓						✓	✓
WS68	A Day in the Life	✓	✓								✓	✓
WS69	Blood Factories				✓					✓	✓	
WS70	Drawing Comparisons			✓	✓						✓	✓
WS71	The Highways of Blood Report	✓	✓		✓						✓	✓
WS72	Blood-Sucking Critters			✓							✓	✓
WS73	Circulatory System Scramble		✓		✓						✓	✓
WS74	Copywork: Manuscript Matt. 22:37		✓							✓		
WS75	Finding Your Place			✓							✓	✓
WS76	Closed vs. Open Circulatory Systems			✓	✓						✓	✓
WS77	Your Beating Heart			✓		✓				✓	✓	
WS78	Pathway of Your Circulating Blood			✓	✓						✓	✓
WS79	On Cruise Control: Along the Circulatory	✓		✓	✓			✓			✓	✓
WS80	Drawing Comparisons			✓	✓						✓	✓
WS81	The Heart			✓	✓							✓
WS82	Getting the Blues: Following the Leader		✓	✓							✓	✓
WS83	Who Am I?		✓	✓						✓	✓	
WS84	Take a Note		✓								✓	✓
WS85	The da Vinci Robot				✓		✓				✓	✓
WS86	Hands as Steady as a Surgeon				✓	✓				✓		

Activity (ACT) / Worksheets (WS)	Description	Multiple Intelligence Type								Ability Level		
		Intrapersonal	Verbal–Linguistic	Logical–Math	Visual–Spacial	Body-Kinesthetic	Musical–Rhythmic	Interpersonal	Naturalist	1	2	3
WS87	Circulating Around/Crunching Numbers			✓								✓
WS88	Heart Health				✓		✓			✓	✓	
WS89	Scrambled Sentences		✓	✓								✓
WS90	Bruise in Sequence			✓	✓						✓	✓
WS91	Solve the Fact			✓						✓	✓	
WS92	You've Got a Lotta Heart		✓	✓							✓	✓
WS93	Got the Smarts Real Good	✓	✓							✓	✓	✓
WS94	Breaking the Ice Idioms		✓	✓							✓	✓

LEVELS K-6
MATH LESSONS FOR A LIVING EDUCATION
A CHARLOTTE MASON FLAVOR TO MATH FOR TODAY'S STUDENT

Level K, Kindergarten
978-1-68344-176-2

Level 1, Grade 1
978-0-89051-923-3

Level 2, Grade 2
978-0-89051-924-0

Level 3, Grade 3
978-0-89051-925-7

Level 4, Grade 4
978-0-89051-926-4

Level 5, Grade 5
978-0-89051-927-1

ATTRACTIVE FULL-COLOR LESSONS

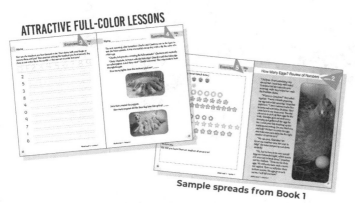

Sample spreads from Book 1

Level 6, Grade 6
978-1-68344-024-6

MASTERBOOKS®
CURRICULUM

AVAILABLE AT

MASTERBOOKS.COM & OTHER PLACES WHERE FINE BOOKS ARE SOLD.

Daily Lesson Plans

WE'VE DONE THE WORK FOR YOU!

PERFORATED & 3-HOLE PUNCHED
FLEXIBLE 180-DAY SCHEDULE
DAILY LIST OF ACTIVITIES
RECORD KEEPING

"THE TEACHER GUIDE MAKES THI
SO MUCH EASIER AND TAKES T
GUESS WORK OUT OF IT FOR

HOMESCHOOL

Master Books® Homeschool Curriculu

Faith-Building Books & Resources
Parent-Friendly Lesson Plans
Biblically-Based Worldview
Affordably Priced

Master Books® is the leading publisher of books and resources based upon a Biblical worldview that points to God as our Crea Now the books you love, from the authors you trust like Ken Ham, Michael Farris, Tommy Mitchell, and many more are available as a homeschool curriculum.

MASTERBOOKS
Where Faith Grows